A WASTING WORLD?

AF735118

A WASTING WORLD?

Natural Resources and Energy

With a Case Study

CURRICULUM DEVELOPMENT UNIT

O'BRIEN

THE CURRICULUM DEVELOPMENT UNIT

The Curriculum Development Unit was established in September 1972. It is sponsored by the City of Dublin Vocational Education Committee and works in co-operation with the School of Education in Trinity College, Dublin, with the approval of the Department of Education. It has a steering committee composed of representatives of these three bodies. This book forms part of the Humanities Curriculum.

Unit Director	:	Anton Trant
Deputy Director	:	Tony Crooks
Humanities Team	:	
Tony Crooks		Coordinator 1972-79
Nora Godwin		1973-79
		Coordinator 1979-
Agnes McMahon		1975-76
Bernard O'Flaherty		1976-78
Dermot Stokes		1977-
Ann Treacy		1978-80

These materials have been edited for publication by Edwin Mernagh and Niamh Mernagh.

Prior to publication, the following schools were involved in the development, use and revision of the collection. The suggestions and comments of the teachers in these schools have been used as a basis for the edition:

Christian Brothers School, James' Street; Coláiste Dhulaigh, Coolock; Coláiste Éanna, Cabra; Coláiste Eoin, Finglas; Coolmine Community School, Clonsilla; Liberties Vocational School, Dublin; Mater Dei Secondary School, Basin Lane; Pobal Scoil Íosa, Malahide; Rosary College, Crumlin; Scoil Íde, Finglas; Vocational School, Ballyfermot; Vocational School for Boys, Clogher Road; Vocational School, Crumlin Road.

First published 1981

O'Brien Educational Ltd.
20 Victoria Road, Rathgar, Dublin 6, Ireland.

ISBN 0 905140 95 8 Hardback
ISBN 0 905140 96 6 Paperback

Jacket Design – Mark Quinn
Typesetting – Lagamage Co. Ltd.
Layout – Redsetter Ltd.
Book Design – Michael O'Brien
Photographs – Eamonn O'Dwyer and others
Printing – Irish Elsevier Printers, Shannon, Ireland.

©Copyright Reserved

All rights reserved. No part of this book may be reproduced or utilised in any form or by any means, electronic or mechanical, including photocopying, recording or by any information storage and retrieval system without permission in writing from the publisher.

ACKNOWLEDGEMENTS

We would like to thank: Patricia Dunphy, Gary Granville, Simon Hewat, Michael King, Seán Lynch, Stan McHugh and Paul O'Sullivan for their contrubution to the development of these materials, and Grainne O'Flynn for reading and commenting on the manuscript. We would also like to thank and acknowledge the following for their help in providing illustrations: Avoca Mines Ltd., Bord na Móna, Delaney McVeigh and Pike, Electricity Supply Board, George Gmelch, Peter Heery, The Irish Times, Mark Quinn and Adrian Slattery.

PLEASE
DONOTTRO
STONESORDIRT
INTHISWA

Contents

Introduction

The earth is a storehouse. It contains countless thousands of substances, used by man for countless purposes. We call these substances Resources, and we depend on them for our very survival. People die if they do not have basic resources such as air, water and food. In our modern way of life, we use other resources to provide clothing and shelter for ourselves, heat for our homes, energy for machinery and fuel for transport. The more comfortable, convenient and sophisticated our lives become, the more resources we use.

Not all of the earth's substances are, as yet, useful to man. To exploit a resource, we must first find it, see it as valuable and know how to make use of it. It is only in this century that uranium was discovered as a useful resource. Previously, uranium was of no practical value because nobody realised how it could be used. Scientists are constantly seeking for new ways in which the earth's substances can be exploited.

AN EVERLASTING STORE?

The world's resources are many and varied; they can be grouped into types and categories, depending on the uses we make of them. Basically, they are of two kinds, those which are "renewable" and those which are "non-renewable":

Renewable Resources: These are substances which, though often used, can be replaced. Some of them remain unchanged by use and are, as far as we know, in endless supply. They include solar energy, wind and wave power, and air. They are called *continuous resources*.

Other renewable substances may increase or decrease depending on the level of usage. Examples of these are soil, water, trees and forms of animal life. Soil contains certain chemicals which are essential for the growth of plants. When the plants die and decay, these chemicals are returned to the soil, which is therefore renewed. If the soil is exhausted by poor farming methods, the renewing process is slowed down or halted, the land can become a worthless desert. These resources which vary with usage are known as *flow resources*.

Non-Renewable Resources: These are substances which have been stored in the earth, often for millions of years. Once used, they are not replaced or renewed. They include oil, coal and metals. Some of them are being replaced but so slowly that man could never hope to exploit this replacement. The world's oil stocks, once exhausted, will take millions of years to renew themselves.

Developments in modern technology have made it possible to locate and exploit the world's resources in a far more efficient way than ever before. Oil has been extracted in ever-increasing quantities since the beginning of this century, and all known stocks will be exhausted in twenty years' time. Young people reading this sentence will only be middle-aged when our oil supplies finally run out, if they continue to be used up at their present rate and if no new

deposits are found. In less than one hundred years we have almost exhausted a resource that took millions of years to form.

Our present way of life, in the rich countries of Europe and America, depends on this exploitation of our earth's riches. Can this way of life continue for ever? Is there, in fact, an everlasting store of natural resources?

It is now generally accepted that certain important resources are in limited supply so their use must be carefully planned. If we do not conserve now, there will be a shortage in the future.

Man has interfered greatly with natural resources and in some cases has damaged them to such an extent that they are no longer useful.

Many areas of fresh water have been polluted by sewage, industrial wastes and fertilisers; in this way an essential resource is needlessly destroyed.

Even in Ireland there are lakes which can no longer support fish, rivers which are no longer fit to swim in and drinking water which has been poisoned and is no longer safe to use.

The pollution of air and of soil has been increasing all over the world. Coal smoke, carbon monoxide from car engines and fumes are already health hazards in many big cities.

It is not enough that we conserve resources which cannot be renewed. Those capable of renewal must also be protected. The survival of mankind depends on the way in which we manage the storehouse of our earth.

MANAGING RESOURCES

How can we ensure that the earth's riches are used in the best and most productive ways? Here is a plan of action:

- We must learn as much as possible about the extent of the resources available to us.
- We must seek to conserve limited resources.
- We must plan the use of resources, eliminate waste, and ensure that poorer countries and future generations have as much right as we have to share in what the earth can provide.
- We must find new ways of providing the energy we need for our way of life. Continuous resources such as solar energy, wind and wave power could be harnessed in order to conserve non-renewable resources such as oil and gas.
- We must change our attitude towards resources. People in rich countries have come to expect a high standard of living. Consider the huge amount of energy needed to run homes, cars and industries. We must stop thinking of our earth as an "everlasting store".
- We must re-cycle materials to get the best possible value from them. This is particularly important in the case of non-renewable resources (such as metals), but also in the case of renewable resources (such as wood-pulp paper) which are in short supply.

THE COST

Until recently, man has regarded the earth as a storehouse of free resources to be used and abused as he pleased. In many ways and in many places, this still happens. Nevertheless, there is a growing understanding that the earth's riches are not entirely free for the taking, there is a cost and it is a heavy one.

A storehouse, as we have seen, needs careful management and it is necessary to see that it does not become empty. It is also necessary to see that careless use of one resource does not spoil others.

The use of oil has had a dramatic effect on the way of life in many countries; oil is a wonderful resource. Our use of oil has led to problems of pollution on a large scale. Even the limitless sea cannot cleanse itself of the thick black sludge from a crude oil spill. Great cities suffer almost constant atmospheric pollution from the fumes and exhaust gases of oil-driven vehicles, factory machinery and heating boilers. Coal has been another great servant of mankind; but mining activities have led to desolate wastelands of slagheaps, abandoned pits and smog.

In recent times, nuclear fuels have been developed to provide alternative energy needed for our modern way of life. Careless use of this resource could lead to disaster. The problem of abuse of natural resources is a worldwide one; Ireland has her part to play. Here are two views of how we manage our natural resources:

IRELAND IS . . .	IRELAND IS . . .
*The last unspoiled land in Europe	*A land whose beauty is marked by random growth of cities and towns
*A land free of air and water pollution	*A land of oil and sewage fouled beaches and seas
*A land of clean beaches, rivers, lakes and forests	*A land contaminated by weed-killers, poisons and powerful fertilisers
*A haven of woods, wetlands and bogs for wild birds and animals	*A land where wildlife has been driven to the most remote corners by overheating, poison, noise and destruction of natural habitats
*A land of balanced urban planning	*A land whose once-beautiful capital city is being torn down in the name of progress

Which description fits? Either may be partly valid today, but both cannot remain true. We must choose. Can we manage to use our resources without damage to the environment? Or are we willing to pay the cost of abuse and live in a dying world? The riches of the earth are there for the taking but we must plan and control their use to make Ireland a better land in which to live, work and play.

RICH WORLD, POOR WORLD

The rate at which resources are used varies greatly from place to place. Most countries in Western Europe have a very high standard of living. The Irish, for example, have a way of life that uses up large amounts of resources. We accept ample supplies of food and water and clothing as normal; good housing, well-stocked shops, and services such as transport, education and health care are regarded as essential. We are constantly seeking to improve the material quality of our lives, by acquiring more possessions and using more resources.

However, resources are not shared equally throughout the world. In the countries of the Third World, most people expect very little from life in terms of material things. Often they consider themselves lucky if they have enough food to eat, something to wear and somewhere to live. They are said to be living at subsistence level. These people use very few resources in their lives.

The basic resources needed for *Life* are air, water and food. As standards of living begin to improve, extra resources are needed. The first requirements are better food and water supplies. Then comes the building of roads, houses, schools, shops and hospitals. This phase of *Development* needs large amounts of resources such as stone (and stone products), wood and metals. Further rises in standards of living place extra demands on the basic resources. It also involves the use of large quantities of *Energy* to fuel the factories and services on which the new way of life depends. This energy can be supplied from fossil fuels, nuclear fuels, wood and the natural sources of wind, wave and solar power. These are, in a very real sense, the key resources of the modern world.

Part One

Resources for Life

Water

The most basic of all resources, those without which nobody can exist, are water, air and food. Almost all of our food is derived from green plants; the animals from which we get meat, milk and eggs are themselves plant eaters. The plants which are the basis of our food supply require fertile land for growth, and so land must be included among the essential resources. In this section we examine the resources necessary for life, and consider the factors which can influence their availability and conservation.

"Psst . . . want some fresh water?"

Water makes up 70% of the earth's surface, but only 3% is fresh water. Less than 1% is available to man for drinking, cooking, bathing and other uses. The rest is locked up in salty oceans, in ice packs or in underground deposits. The amount of water in the world has not changed for thousands of years.

Water cannot be destroyed; it can be converted to steam or ice or it can become dirty or unusable, but it is still water. In this way it is a renewable resource. Where does water come from? When it rains or there is a fall of snow, sleet or hail, the water soaks into the ground or runs off it. It then forms streams which join rivers and lakes, and eventually reaches the sea. Water from the surface is constantly evaporated. Plants transpire water into the atmosphere. In these ways, water vapour rises and forms clouds. These condense, and the water returns to earth as rain, snow, sleet or hail. This unending process is called the *Water Cycle*.

THE WATER CYCLE

We cannot live without water. Our bodies are 65% water, and we depend on it for almost everything we do.

The different ways we use water include:

- Drinking and washing in the home
- Carrying away sewage and other waste
- Industrial processes
- Producing power
- Irrigating land
- Transporting people and goods
- Fishing
- Leisure pursuits

An ample water supply in the home is often taken for granted nowadays; but how does the water get to us? Rainwater is collected from streams and rivers into reservoirs, often a natural lake is used. This water is then cleaned and purified to make it

THE WATER CYCLE

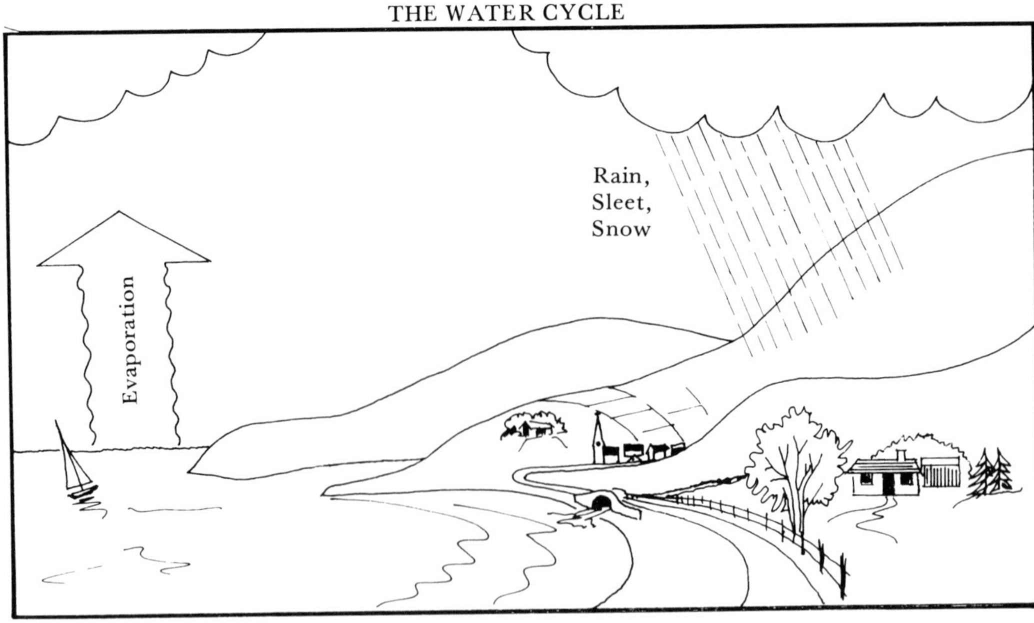

suitable for people to use. First it is pumped into giant storage tanks and allowed to settle. Chemicals are added in a chlorination plant to ensure that harmful germs and bacteria do not contaminate the water. Finally it is pumped into a supply reservoir, which in turn feeds the main pipes from which our domestic supply is taken.

HOW MAN USES WATER

Water in the Home

Average amount of water used in the home:

We drink each day:	1½ litres
Taking a shower:	70 litres
Flushing a toilet	9 litres
"Twin-tub" automatic washing machine	45 litres

When water is easily available we use it freely. When it is not available, we can manage on very little. In rural districts with no piped water supply, only about 20 litres are used per person each day. It is estimated that in 30 years' time we shall be using twice as much water as at present.

Water in Industry

Most water in industry is used for cooling. Power stations and chemical industries, for instance, use large quantities for this purpose. Water is pumped out of a nearby river, used for cooling and then in some cases is returned to the river. If this water is still warm when it is returned, it has a destructive effect on the fish and plant life of the river. However, many factories have closed cooling systems and re-use the same water again and again. This is a good example of the re-cycling of a resource.

As well as using water for cooling, some industries also use water in their products, for example breweries. Water is also used as a source of power in industry. In recent years it has been used to generate electricity. This can be done in a number of ways. The method most commonly used is as follows:

A HOUSEHOLD WATER SUPPLY

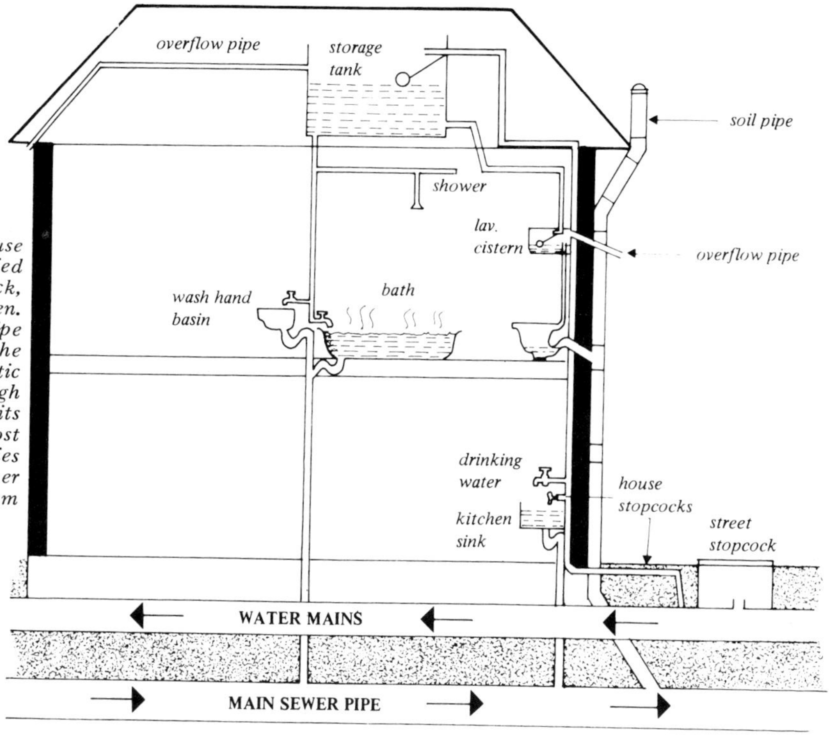

Water entering the house from the mains is carried to a tap or stopcock, usually in the kitchen. From there, a branch pipe takes the water up to the storage tank in the attic (mains water is under high pressure, and will rise of its own accord). In most houses, this tank supplies water to all the other outlets, toilet, bathroom and kitchen sink.

How water gets from reservoir to the house.

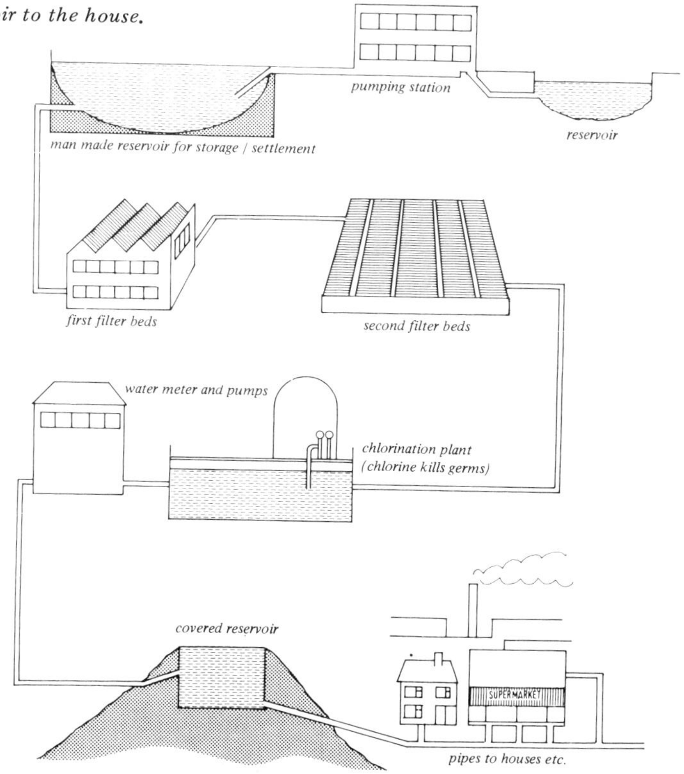

WATER IN INDUSTRY

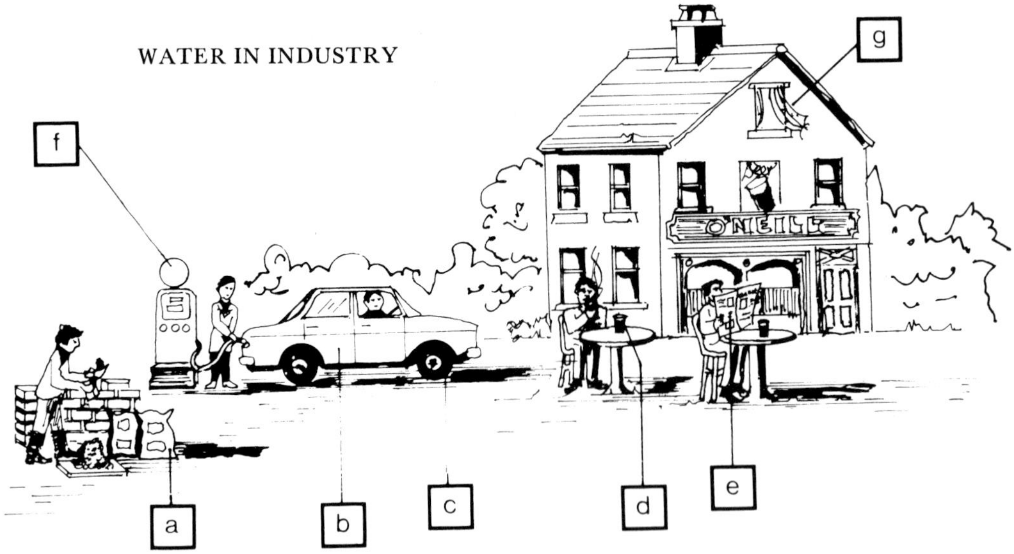

(a) 180 litres is used in making a bag of cement.
(b) A four-door family-size car requires over 450,000 litres to manufacture.
(c) To make one tyre requires 190,000 litres.
(d) 350 litres of water is used in making one litre of beer.
(e) 190 litres is used in making the paper for each copy of a newspaper.
(f) One litre of petrol takes over 70 litres to refine.
(g) One kilogramme of nylon requires 1,000 litres in manufacture.

A suitable place on a river is chosen where a dam can be built to make an artificial waterfall. Water tumbling down the falls turns turbines connected to a generator, and thus produces electricity.

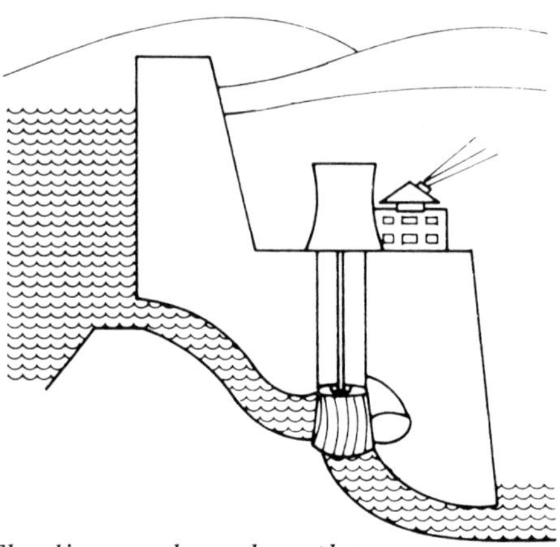

The diagram shows how the force of the water turns the turbines which generate electricity.

There are 10 E.S.B. hydro-electric stations in Ireland, situated on the rivers Shannon, Liffey, Erne, Clady, Lee, and at Turlough Hill, Co. Wicklow.

Water on the Farm

Agriculture makes a great demand on water. To grow the daily food for one person in this country nearly 10,000 litres of water are needed. In Ireland, because of the mild climate, there is plenty of water for growing crops and feeding our livestock. Farmers in desert areas have to face the problem of receiving very little rain per year. To overcome this problem, they have to irrigate their land. Irrigation is the artificial watering of land in order to overcome shortages in rainfall.

About 200 million hectares of the world's agricultural land is irrigated. By means of irrigation, the farmer can get a high yield from his crops even if he lives in a very dry area.

HYDRO-ELECTRIC STATIONS IN IRELAND

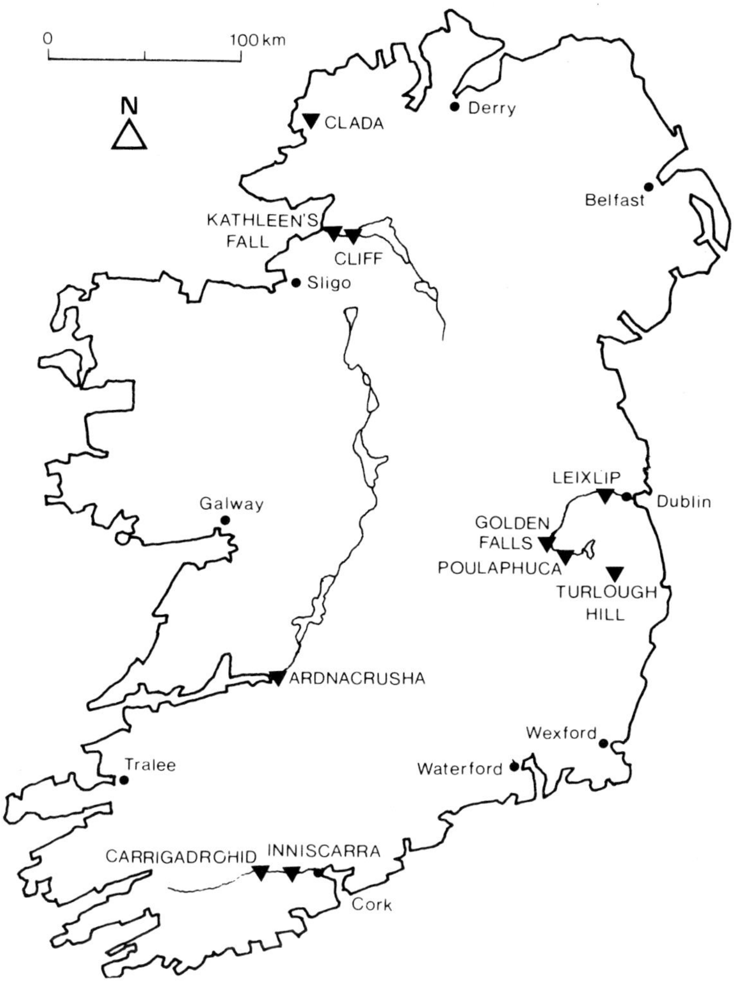

WATER POLLUTION

The water we drink is clear, colourless and odourless. This is because it has been purified. Water in its natural state is never pure. Many substances are contained in it, which are necessary for the plant, fish and other animal life which is to be found in water.

Water is only termed *polluted* when it contains substances harmful to life. A polluted river has a smaller variety of plant and animal life than normal or possibly none at all.

Wastes dumped in water can pollute. To a certain degree, rivers can dilute and break down this waste, but if they are overloaded, they lose this ability and become heavily polluted. The rivers then carry this pollution to lakes and to the sea.

Wastes which cause pollution in our rivers and lakes include the following materials:

- untreated or partially treated sewage
- industrial waste which often contains harmful chemicals
- rubbish dumped into rivers and lakes or left on beaches
- farm wastes such as silage effluent,

manure and chemicals from pesticides and artificial fertilisers. Some of these wastes are rich in nutrients which cause water weeds and algae to grow very rapidly. This plant life uses up all the oxygen and leaves none for the fish and other animal life

- oil, spilled from tankers, is the most serious source of pollution in the open sea
- heated water from industrial cooling plants which can raise the temperature of the water and upset the balance of the natural habitat of its animal and plant life

Examples of Water Pollution

Minamata, Japan

Minamata, on the southern Japanese island of Kyushu, is a farming and fishing village. In 1907, a chemical company, Chisse, decided to set up a factory there. The people were eager for the wealth and prosperity this would bring to their village. All the mercury-laden chemical wastes from the factory was dumped into the sea around Minamata. As these wastes increased, the level of mercury in the sea increased and found its way into the plants and animals there. Fish was a major part of the diet of the people. Soon Minimata was struck by a strange disease. Many people died suddenly, others were confined to bed for years and eventually died and some people became disfigured and were unable to control their limbs. This disease was caused by mercury poisoning contracted from mercury infected fish.

Chisse, the firm responsible for this disaster, paid compensation money to victims and relatives. This was of little consolation to the people of Minimata whose lives were destroyed

Polluted stretch of water on a canal.

"Now maybe they'll be moved to do something about Water Pollution!"

by a tragic event which could have been avoided if the firm had acted in a responsible way.

Lough Ennell, Co. Westmeath

Lough Ennell was once a clear fresh water lake enjoyed by tourists and fishermen alike. The lake is close to the town of Mullingar, and sewage from the town flows into it. As the population of Mullingar increased, so also did the amount of untreated sewage entering the lake. In addition, slurries from pig farms in the area are allowed to seep into Lough Ennell. These slurries are rich in nutrients.

The discharge into the lake has increased its growth of algae. These plants use up all the oxygen, leaving none for the fish and other animal life. If this continues, eventually the lake will smell like a piggery or sewage pit, and the surface will be covered with dead and decaying weeds. It will be a dead pool of polluted water. In this way a fresh water source and a tourist attraction will be destroyed.

Why Must Water be Protected?

Water is a basic resource needed for life. If it becomes polluted it poses a grave threat to man's existence.

- Polluted water is a danger to health. Cholera, a disease caused by contaminated water, claims tens of thousands of lives every year. Many other diseases, too are caused by polluted water.
- It is harmful to water-dwelling animals and plants. Fish, for example, cannot survive in heavily polluted water. Pollution upsets

the natural balance of animal and plant life in the water. Diseased fish can cause serious disease.

- The cleaning of polluted water to make it safe for household use is a complicated and expensive process.
- Polluted rivers flowing into the sea build up banks of sludge which are a hindrance to shipping.
- People like to use rivers for swimming, sailing and other pursuits. They cannot use them if they are polluted.

WATER SHORTAGE

We have seen how essential water is to man's existence and way of life. We have seen how it can be damaged by pollution and misuse. There are some parts of the world, however, which suffer from a simple shortage of this most essential basic resource.

Climate is the most important factor in the availability of water. Ireland has a "temperate" climate; the main features of this are the absence of extremes of temperature (it is never very hot or very cold) and a fairly high rainfall (763 mm to 1500 mm per annum). This climate provides our country with ample and reliable water supplies. Local shortages are usually due to inadequate piping, pumping or reservoir facilities.

DESERTS

The world rainfall map shows the areas which have less than 250 mm of rain per annum. Some of these are situated in the far north, around and beyond the Arctic Circle. The rest are desert areas, which show common features in arid, barren landscape, thin soil and sparse plant

WORLD RAINFALL

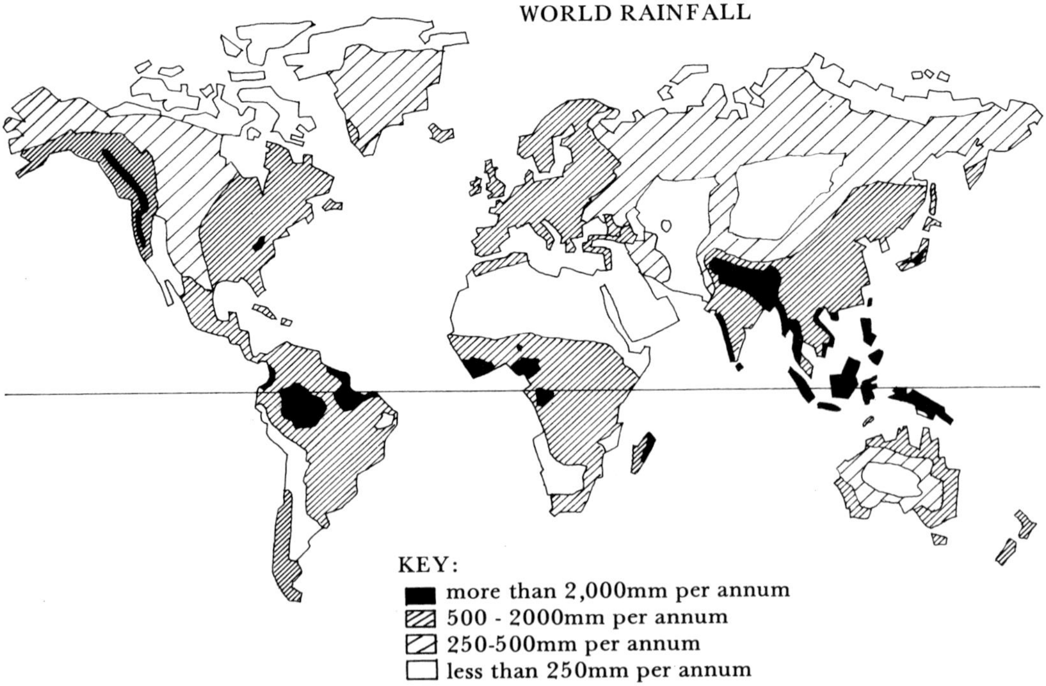

and animal life. In many deserts daytime temperatures are very high.

Very few people live in the majority of these places, as the absence of water makes life extremely difficult. Some desert lands, however, have valuable natural resources that can be exploited. Saudi Arabia, the biggest oil producer in the world, is a desert. In such cases, elaborate and expensive measures have to be taken to provide water for domestic uses, to irrigate fields for cultivation and to fulfil the needs of industry. These measures include deep boring for wells, long-distance piping, desalination (removing the salt from sea-water to make it usable), and even transporting water in tankers.

DROUGHT

Many parts of the world experience occasional droughts, this happens where there is a shortage of water. The areas shown on the map with an average annual rainfall between 250 mm and 500 mm e.g. Saudi Arabia, Central Australia, suffer in this way. While these areas can support large populations, the periodic shortage of water is a great obstacle for them in the development of a high standard of living.

"The United Nations Secretary-General, Dr. Kurt Waldheim, shocked by the disaster in the Sahelian zone of West Africa, where famine and disease killed an estimated 100,000 people last year, has warned the worst is yet to come.

Speaking about a recent visit to the drought-ravaged area, he said in New York last week: 'I have seen with my own eyes the toll in human suffering, the depletion of resources and the setback to development plans.'

It would take 'a bold international effort of enormous scope both to mitigate the present suffering and to provide for a better future,' Dr. Waldheim said . . .

The region covers six countries—Senegal, Mauritania, Mali, Upper Volta, Niger and Chad—with a population exceeding 20 million. The drought directly affected about 2 million people, most of them nomadic herdsmen."

The Irish Times, 11 March 1974

Air

Like water, air is a renewable resource. It is used, "cleaned" and returned for use again in an everlasting cycle. Like water, it too can be spoiled by pollution. Unlike water, air is ever-present; there are no "local shortages", or "seasonal shortages", unless we consider the fact that air is "thinner" (that is has less oxygen) at high altitudes. For all practical purposes, an adequate air supply can be found anywhere on the surface of the earth.

What is air?

Pure air has no colour, taste or smell. The air we breathe is a mixture of several gases, mainly oxygen and nitrogen. Oxygen is the "essential" ingredient in the air which is vital to man's existence. The earth is completely surrounded by a 60 kilometre thick gaseous layer, called the atmosphere.

What happens when we breathe?

When air is taken into the lungs, an exchange takes place. The lungs take the oxygen from the air, to be circulated around the body in the bloodstream. At the same time, a waste product called Carbon Dioxide is expelled. A similar exchange takes place when anything is burned. Oxygen from the air is used in the burning process. The air we breathe always contains some carbon dioxide.

How is air "cleaned"?

Green plants complete the "air cycle" that provides animal life with its vital oxygen supply. Plants take in carbon dioxide, which man and animals breathe out as a waste product. This gas is used by the plant in a process called photosynthesis to provide carbon for growth. The plants then "breathe out" oxygen as a waste product.

Therefore a cycle is set up whereby man breathes in oxygen and breathes out carbon dioxide, and plants take in carbon dioxide and give out oxygen. In this way there is a constant supply of oxygen and carbon dioxide. This cycle regulates itself, happens automatically in nature and like the water cycle, is continuous.

AIR POLLUTION

Air can be contaminated by many means: smoke, from chimneys, industrial fumes, dust and grit from quarries, exhaust fumes from cars, trucks and buses, weedkiller sprays, insecticide sprays.

Polluted air can cause health problems (including cancer) in humans and animals; it damages and even kills plants, including trees; even stone can be attacked by polluted air. The surfaces of many stone buildings in

Dublin have been worn and pitted by acid chemicals in the air.

Because Ireland is a thinly populated country with relatively little industry, it has not suffered massive air pollution such as that in the heavily industrialised areas of California in America or the Ruhr Valley in Germany. Nevertheless, we must be concerned about the quality of the air we breathe. Air is an international resource, as it is never still. Winds can take a mass of polluted air and sweep it along so that it affects the lives of people hundreds of kilometres away. Luckily, most of our winds come from the Atlantic; this shelters Ireland from much of the industrial smog of Britain and the industrial centres of Europe.

Land

All living things need nourishment to survive. Plants require soil, air and

THE OXYGEN CYCLE

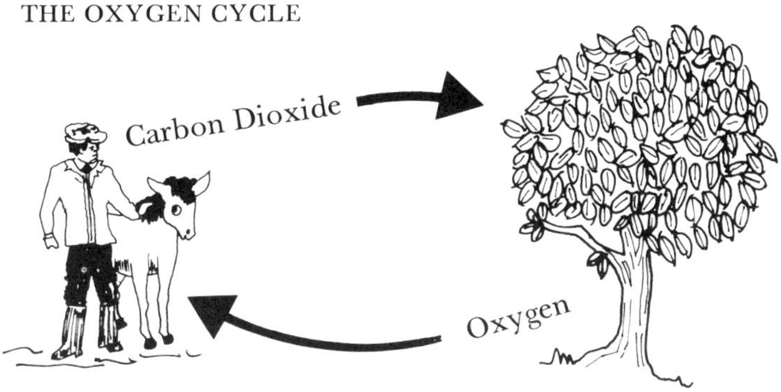

Below: A FOOD CHAIN

1. *Green plants are the only living things that make their own food. Carbon Dioxide in the air combines with water in the leaf to make sugar.*
2. *When insects and animals eat green plants, they absorb the nutrients and energy in them.*
3. *When insects and small animals are eaten by birds and larger animals the nutrients and energy are passed on.*
4. *Man absorbs the nutrients and energy of green plants directly by eating vegetables and cereals, and indirectly by eating meat.* *a*

A FOOD CHAIN

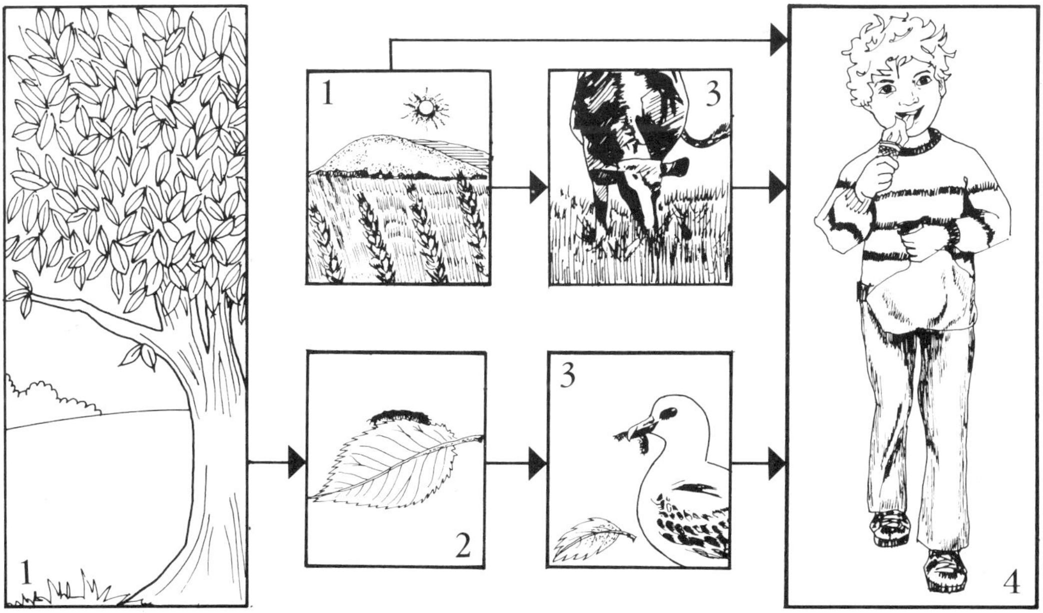

the sun. Animals feed on plants insects and other animals. Man's nourishment is provided by plants and animals of the land and sea. All man's sources of nourishment depend ultimately on plant life. The animals from which we derive food such as meat, eggs and milk products are themselves plant-eaters; green plants are therefore the basic food resource the earth provides.

Everything we eat, therefore, depends on green plants. The growth of plants depends on energy from the sun and on the fertility of the land. In a sense, after air and water, land is the most basic resource of all.

SOIL

Soil is formed when the action of rain, frost and sun causes rocks to decay. Insects and the roots of plants push into the cracks in the rocks and split off small fragments. Over thousands of years these fragments break down into very fine particles. These particles, together with dead plant material, become soil.

Plants which grow in the soil help to protect it. The plant's roots bind the soil together and decaying leaves form a dark layer on top of the soil. This layer, known as humus, helps to stop the soil from drying out. If the plants are removed and the humus layer is broken up by cultivation, the soil is not protected. Wind and rain can wear away, or erode, the soil. This has happened in many places in the world, including parts of the Aran Islands. The blanket of soil covering the earth's surface is very thin. In some areas it is six metres deep, but it is often less than a metre or only a few centimetres deep. The soil provides plants with most of the nutrients or food they need. Since crops cannot grow without nutrients, a good soil built up over

THE NUTRIENT CYCLE

1. *Nutrients enter the soil in rainfall.*
2. *Other nutrients enter the soil slowly from the gradual breaking down of rocks. These chemicals are used by plants (in this case, a tree) to provide the nourishment needed for growth.*
3. *Worms, insects, fungi and microscopic organisms break down leaves and branches and thus return the nutrients to the soil. When the tree dies, it is recycled in the same say.*
4. *Only a few nutrients are washed away by the rain because the tree protects the soil by shelter and by the binding action of the roots.*

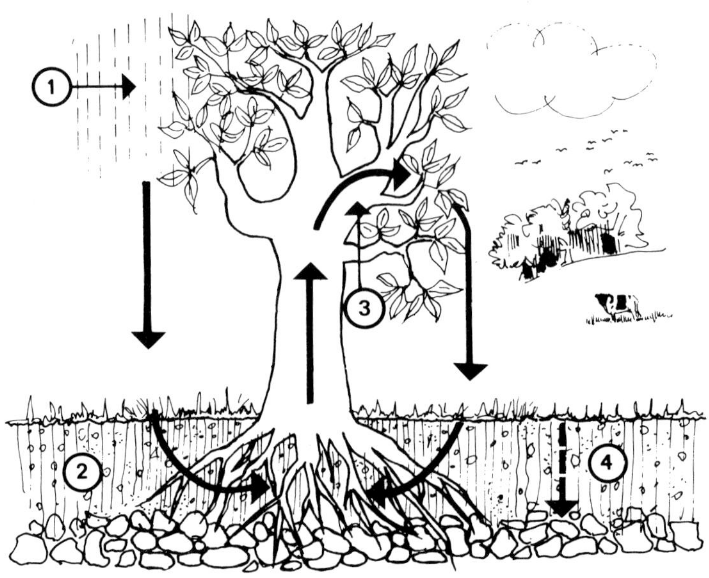

thousands of years is very precious.

Nature has been "re-cycling" resources for millions of years. This diagram shows how, in an undisturbed natural environment, the same chemicals are used again and again to promote the growth of generations of living things, through the build-up of a healthy soil.

When land is cleared for farming, the nutrient cycle is broken down. Nutrients are no longer returned to the soil by decaying vegetation, and without the cover of trees to protect the soil further nutrients are lost through the action of wind, rain and sun. The soil quickly becomes "exhausted" unless the farmer replaces the lost nutrients by fertilisers.

EROSION

With rising populations, more and more land is needed for growing food. Unfortunately, there are many problems in making new farming land. Erosion, or soil loss, is a very serious problem in many parts of the world.

These diagrams show how it can be caused by bad farming methods:

1. The soil has developed over thousands of years in the forest.
   Then the farmer cuts the trees down.
   The nutrient cycle that formed the soil is broken down.

2. He now plants crops on the hillside. Ploughing breaks the soil down and nutrients leave the soil each year in the harvest. The soil becomes less porous and can't soak up the same amount of water. Rainwater therefore runs down the slope, taking some of the topsoil with it.

3. The soil is now too poor for crops so it is used for pasture. If the soil is grazed heavily, bare patches appear. Rain runs down the animal tracks taking soil with it. When the pasture is no longer rich enough for cattle, it is used for sheep or goats.

4. The sheep have stripped the remaining pasture. Rainwater has carved the animal tracks into gullies which become deeper as the rain washes more soil down them. In the end, all the topsoil is washed away, leaving the rock bare. This fertile hillside has become a barren desert.

MAN-MADE DESERTS

Some parts of the world are natural desert; they lack the basic climatic and soil conditions to sustain plant life. The extreme cold conditions of Antartica and the aridity of the Himalayas has only supported pioneer vegetation. In contrast, many of the places we know as deserts today, the Sahara for example, were once fertile. A recent example of man's destruction of land as a natural resource is the case of what has come to be known as "The Dust Bowl" in the American mid-west. The State of Oklahoma was famous as a fertile grassland, providing grazing for great herds of cattle. Then farmers ploughed up the grass to plant crops; this disturbed the thick masses of grass roots that held the soil in place and kept the water in the earth.

In the 1930's there came a long period of drought, and the soil dried up. This was followed by great sandstorms which blew the soil away. Dust and lack of water destroyed what was left of the crops, and soon the land was covered in a thick blanket of bare, dusty soil. The disaster area was aptly named "The Dust Bowl". Most of the farmers left their land, which was now ruined and useless.

It has taken many years, and millions of dollars from the United States government, to rescue the land and bring it back into production. The Dust Bowl is a grim reminder that all resources, even the land, need careful management and conservation.

Part Two

Resources for Development

Metals

Metals are a vitally important resource. We all use metal objects in our daily lives; much of the technology needed for our modern way of life is based on machines, tools and appliances made of metal. There are many different metals and alloys (that is, combinations of metals) with many different properties and uses, but all have one thing in common; they are a non-renewable, limited resource. The copper deposits in the world will eventually be found and used up, except for deposits which are too difficult to get at. When this happens, copper will cease to be a resource, it will never be replaced.

For a long time, it has been known that the deposits of metals in the world were limited. However, it is only in relatively recent times that this has become a major problem. Scientists have calculated that many of the metals we so commonly use may be all used up within a couple of generations.

METAL	ESTIMATED LIFE FROM 1980
Aluminium	100 years
Iron	240 years
Copper	35 years
Zinc	22 years
Tin	15 years
Lead	25 years

(US Bureau of Mines)

These figures are only estimates. They do not take into account that new deposits of these metals may be found. However, our metal resources will run out sooner or later and they

PRINCIPAL WORLD METAL RESOURCES

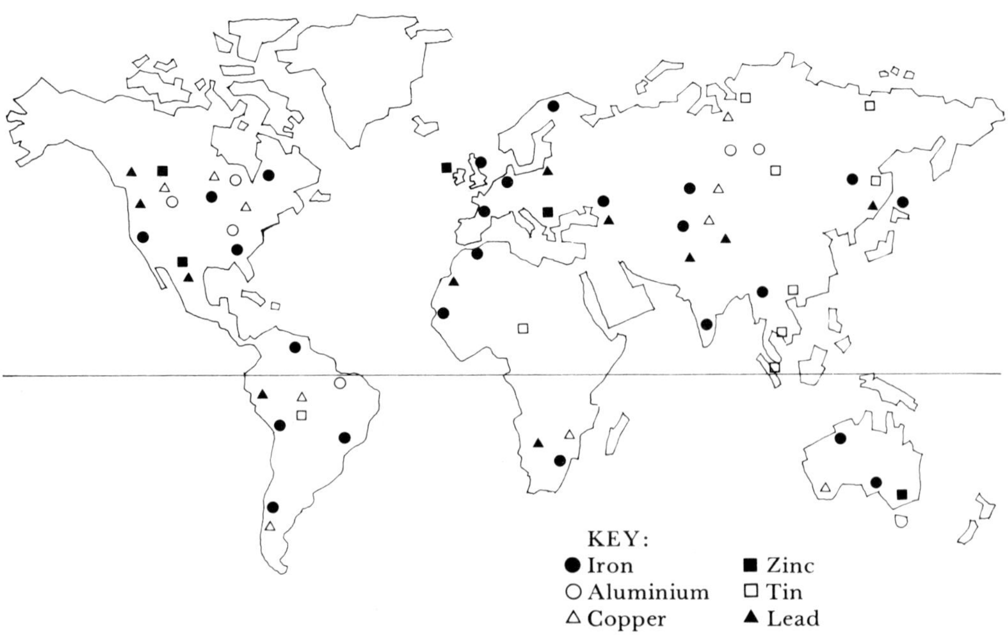

cannot be replaced. This will have a great effect on our present way of life.

The most important metals used in the world today are: Iron, Aluminium, Copper, Zinc, Tin and Lead.

Iron: Iron is perhaps the world's most important metal. From iron we make steel, and from steel we make many of the things we use every day: buildings, cars, ships, trains, tools, machinery of all kinds. A large part of the world is composed of iron, the earth's central core is probably pure iron. It never occurs in its pure state on the surface where man can extract it easily. Iron is found with other substances in a mixture called *ore* and the ore itself is embedded in the rock that forms the earth's crust. Extracting it is a long, difficult and expensive process. When iron ore lies near the surface it is scraped out of shallow pits by huge machines. If it lies deep underground, it is mined out like coal. Most of the world's iron ore is converted to steel. Using chromium as an alloy, stainless steel is produced, it is extremely hard and resists corrosion.

Aluminium: There is more aluminium than iron in the earth's crust, but most of it is contained in hard rocks from which it cannot be extracted. In some places, these rocks have been broken down and weathered into a kind of clay called *bauxite*. It is from this clay that we get the aluminium used in modern industry. It is a light, strong metal, does not rust or easily corrode and conducts electricity. It is used in the manufacture of aeroplanes, engines, many household appliances, and scientific instruments.

Copper: Copper is one of our most useful soft metals. An excellent conductor of electricity, it is widely used in electrical equipment and as a roofing material. Combined with zinc it becomes brass; when tin is added, it becomes bronze.

Zinc: This metal is seldom seen in its pure state. It is used mainly as a coating over steel to prevent rust, and alloyed with copper it forms brass.

Tin: This soft light metal was originally used in the manufacture of bronze. Since the discovery that tin-cans have special qualities in preserving food, it is widely used in the food processing industry.

Lead: Lead is not as important now as it once was, but it is still used in the building industry for piping and as a roofing material. It is soft and heavy.

SAVING OUR METAL RESOURCES

Recycling

Recycling of resources means using a resource again in a different way. Old cars, fridges and other metal objects can be used to gain scrap metal which is then processed to make new metal goods. Recycling lessens the drain on the world's metal resources.

Using Metal substitutes

Often metal is used for purposes where it is not essential, and because of this, a valuable resource is wasted. Many of our metal goods are purposely built for a very short life. Cars, for example, go "out of date" in a very short time, spare parts for a particular model become unavailable and often it is necessary to buy a new one. This leads to an ever-increasing use of metals for replacements.

By using synthetic products like plastic, we can preserve more metals, but this alone will not solve the problem. Today, crude oil distillates, from which plastics are made, are becoming less plentiful and more

PRINCIPAL METAL MINES IN IRELAND

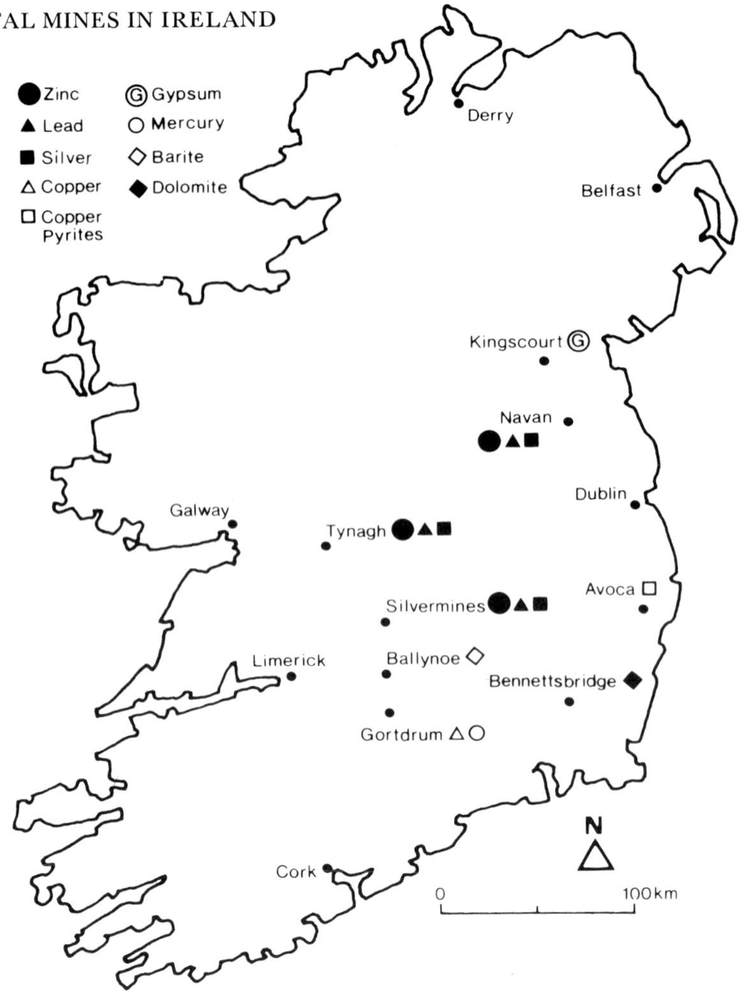

costly. This has led to plastic price increases and shortages. Another serious handicap is the problem of the usage of plastic scrap. Light plastics can be shredded and recycled but hard plastics are difficult to recycle. The great bulk of plastics are buried, burned or dumped in the sea. If large amounts of these wastes are dumped at sea, they lead to water pollution.

Using Other Metals

There are seventy different types of metals. As the supplies of the six major metals run out, it will become necessary to develop ways of using other metals.

MINING IN IRELAND

"The old mine is situated in the lowland of Carloon and about a mile and a half N.E. of the village of Tynagh . . . That in former times it was worked extensively is evident from remains of old shafts . . .the inhabitants of the hamlet informed us that when the grounds are tilled in Spring all the fowl die, being poisoned by something that they pick up . . ."

(Memoir quoted in the Geographical Survey of Ireland 1965)

For years it was believed that Ireland was not richly endowed with metals and other minerals. Yet Ireland has a tradition of mining going back to the Bronze Age. The great bulk of the country's mineral resources were unrecognised. They could not be properly developed until the right level of technical knowledge and expertise became available.

The early 1960's saw the rapid development of the Irish mining industry. Licenses for mineral prospecting were granted to many mining companies mainly North American in origin. There are eight major mining operations in Ireland today. The location of six were found directly by government bodies or were known to exist for centuries.

The government felt that it could not pay the cost of developing the mines. It encouraged outside mining companies to carry out the work instead by offering them tax-free concessions. In return for their initial outlay on plant and machinery, they would be guaranteed a high profit when the minerals were sold on the world market. Ireland now has the largest lead and zinc mine in the world at Navan, the largest underground zinc mine in Europe at Silvermines, and, the largest lead producing mine in Europe at Tynagh. The Navan mine is the largest of these resources, two companies are involved in working this deposit: Tara Mines Ltd. and Bula Ltd.

Tara Mines, the larger of these companies, was formed to exploit a lead and zinc deposit discovered at Navan, Co Meath in 1970. At full production, it will provide 470,000 tonnes of metal ore per year. The ore is shipped from Mornington, near Drogheda. It has to be transported abroad so that the metal can be extracted, as Ireland has no smelter. Smelting is the process which extracts the metal from the ore.

MINE	PRODUCTS	WHO FOUND IT
NAVAN	Lead, Zinc, Silver	Soil Division of the Agricultural Institute
TYNAGH*	Lead, Zinc, Silver	Irish Geological Survey
AVOCA	Copper Pyrite	First discovered about 1730
BENNETSBRIDGE	Dolomites (Magnesium Limestone)	Irish Geological Survey
BALLYNOE	Barytes	First discovered in 1860's
SILVERMINES	Lead, Zinc, Silver	History of mining dating back to 1604 and earlier

**The mine at Tynagh, Co. Galway is being closed down, as most of the good-quality ore has been extracted and it is considered uneconomical to continue working the mine.*

The ore body at Navan extends from just underground to known depths of at least 1,700 metres. Tara Mines use underground shafts to extract the ore. Besides the vast tunnel complex, the mine consists of various facilities for processing the lead and zinc at the rate of 7500 tonnes of ore per day. These facilities include storage buildings, mills to crush the coarse ore and concentrators to remove waste materials. There are also extensive buildings housing offices, laboratories, training centre, warehouses and workshops and worker facilities.

But the mines at Navan raise an important question: should the natural resources of a country be owned by private companies?

As a shareholder in the Navan mining companies, the Irish Government has only part ownership of them (25% of Tara and 49% of Bula) and, with the other shareholders, gets its appropriate share of the profits. Many people feel that this is not enough and that, instead, the State should own and control Ireland's mineral wealth.

From Mine to Smelter

There are three basic steps in the conversion of a base metal from an ore deposit in the ground to the finished product. They are:

Extracting stage
Smelting stage
Processing stage

During the first stage, the ore is mined from the ground, it is then crushed and prepared for transportation. This stage is the least complicated and the least expensive to operate. Once initial costs have been met (acquiring land, moving in plant and machinery, laying essential services, roads, water supply and electricity) the return to the mining company of profits from the exported ore is impressive. The Tynagh mine, for example, was brought into production at a cost of £41 million and the profit made in the years 1965-1970 was nearly £20 million. During this period, an average of 300 mine workers were employed and the wage bill was £2.6 million over this five year period.

Stage two involves the smelting or extracting of pure metal from the ore, thereby doubling its value. Ireland has not yet built a smelter. At present, untreated ore is shipped to smelters on the Continent. The metal is then processed further into a whole range of finished goods. (Stage three.) This increases by tenfold the value added to the metal. As yet there are no facilities for processing native metal resources in Ireland.

As we have seen, Irish metals have to be exported for smelting and processing, therefore a major part of the profit is lost.

The Irish Government has considered building a smelter. The costs are high, but jobs would be created not only at the smelter but also in metal processing industries. The American Mining Corporation was invited to participate in building the smelter but in 1978, they opted out of the project. The reason given was the failing world demands for zinc products. The Government of the USSR has offered to build a smelter and the Industrial Development Authority is examining the matter.

The site for the proposed smelter will have to be carefully chosen. There will be large amounts of toxic waste to be disposed of. The location will probably have to be by the sea, Cork Harbour has been suggested as a possible site. The smelter must have strict regulations governing how it disposes of its effluent. The waste will include millions of litres of water that has been used for cooling and purifying, and sulphur dioxide gases emitted from the furnaces. Although the building of a smelter may be vital for our economy, it must not be done at the expense of the environment.

Other Minerals

Apart from base metals, two minerals of special industrial importance are available in large amounts in Ireland, barytes and gypsum.

Barytes, a white substance, is used in the manufacture of paint and rubber products. In its less pure forms, it is made into the heavy "mud" which is used in oil-drilling operations. In Ireland, barytes have been mined by Magcobar Ltd at Ballynoe, Co Tipperary, since 1963. This material is ground into drilling mud and exported to oil exploration sites all over the world. In Co Sligo, pure barytes were also mined by Benbullen Barytes Ltd between 1943 and 1960, and a new company plans to re-open production there.

Gypsum, a reddish substance, is used in the manufacture of plaster and plaster board. Ireland's only known deposit of gypsum is at Kingscourt, Co Cavan, where Gypsum Industries produce about 400,000 tonnes per year.

The Mining Industry in Ireland now directly employs 2,500 people. The real wealth from mining lies in the creation of mineral-using industries. To date we have no such industries, and as a result, the real value of Irish mines is being exported. We will only obtain the true value of our resources when we have the capacity

A gravel quarry at Blessington, Co. Wicklow.

to process the raw material and market it ourselves.

Wood

Wood, or timber, is unique among the construction resources in that it is renewable. A forest can be planted, grown to maturity, felled and planted again in an unending cycle. Unfortunately, there is one great limiting factor in the way we can exploit this resource: trees take a long time to grow, so the renewal span for a forest is long from 50 to 200 years, depending on the type of tree. Overproduction of timber from one area can result, therefore, in the clearing of the whole woodland. Our own history gives us an example of this.

Long ago, most of Ireland was covered in forests of oak, ash, elm and birch. Over the years, these trees were cut down to clear the land for farming and to provide timber for building and fuel. Great amounts of timber were sold abroad; to England for shipbuilding, to Wales for pit-props in the mines and to France and Spain for making wine barrels. By the beginning of this century, the forests were gone, with the exception of scattered patches. Even today, after fifty years of planting by the Government Forest and Wildlife Service, Ireland is the least-wooded country in Western Europe. If we look at these percentages within the E.E.C. and compare them with ours, we get an idea of how little forestry we have.

Luxembourg	31%
West Germany	29%
Italy	21%
France	20%
Belgium	19%
Denmark	9%
Netherlands	7%
Britain	7%
Ireland	3½%

Tree felling in Avondale Forest, Co. Wicklow.

Wood is used for many purposes. A single house contains a huge quantity of timber—rafters, joists, floors, windows, doors and most of our furniture. Paper and cardboard are often made from pulped timber, and as these materials have a short life, (newspapers and empty cartons are soon discarded) the amount consumed is large. Finally, wood is used as a fuel for burning.

Most of the 3½% of land under forest in Ireland has been planted over the last fifty years, mainly on mountain, bog or other marginal land where trees are the only crop that will grow. The trees are almost all conifers, species such as spruce, pine, fir and larch. Conifers are not native to Ireland, but they are planted in preference to the native hardwood species as they can survive on poor land and grow quickly. They provide useful cuttings in 12 to 15 years and are ready for felling within 50 years of planting and provide a clean, straight trunk for easy, waste-free timber production.

Forests grow well in Ireland because of our moist climate and freedom from extremes of temperature. In fact, some tree species grow at a faster rate in Ireland than they do anywhere else in Europe. Yields of Norway Spruce, for instance, can be three times greater in Ireland than in Scandinavia, while a Sitka Spruce forest can grow at an even faster rate. At present about 28,000 hectares are being planted every year throughout Ireland. Expressed in terms of the population this figure means that 12 trees are planted annually for every man, woman and child. Even at this rate of planting, there will be less than 1½ million hectares of forest in Ireland by the end of this century or less than 10% of the total land area. Young trees are reared in a nursery and planted out after about three years. After 10 to 15 years of growth, and at intervals from then on, the forest is "thinned" that is, the less healthy trees are cut out, leaving the best specimens to grow to maturity. Thinnings are used to make wood-pulp in the paper mills, and in the manufacture of chipboard and wall board. Mature timber is sawn into planks for building and construction use.

Ireland imports millions of pounds' worth of timber annually, mostly from Canada and Finland. As our own forests reach maturity, more and more of our needs can be supplied from our own resources. Apart from the production of timber, woodlands themselves have value and importance. They provide shelter from the wind and help to prevent soil erosion in places where few other crops would grow. They provide habitats for many forms of plant and animal life, some of which might otherwise be driven to the point of extinction. Finally, woodlands are a wonderful amenity, providing recreation for thousands of people in the form of forest parks, picnic sites, forest walks, orienteering events, opportunities for bird and animal spotting. The forest is one natural resource whose use makes a really positive contribution to the environment and the quality of life in general.

Stone

There is no shortage of stone; it is a non-renewable substance, and exists in great abundance on the earth. Stone has always been one of man's favourite building materials, except in areas where timber is very easily available. Stone buildings have long been a feature of the landscape in most European cities. Nowadays, few buildings are made of stone in its raw form; most are constructed of concrete blocks. This material combines the strength of stone with the convenience of standard shape, size and texture. Concrete, cement, sand and gravel are all products using the same basic raw material, stone.

In Ireland, quarrying still takes place for building stone; granite quarries are worked in Co. Wicklow, quartz stone is cut in Co. Donegal. Roofing slate was once produced on Valentia Island, and flagstones at Liscannor, Co. Clare. Far more important commercially nowadays are the sand, gravel and crushed stone that are used in the construction of buildings and roads. Limestone is the most common material for this purpose. There is certainly no shortage of available rock. Much of the material needed is extracted from sand and gravel pits rather than quarries. There are many of these all over Ireland, often availing of the glacial material deposited during the ice age in ridges called eskers and moraines.

Stone, or limestone to be more exact, is essential to the cement industry. There are three cement factories in Ireland: two at Drogheda and one in Limerick. Between them, they make 1.5 million tonnes of cement per year from local limestone, shale or clay, and from gypsum at Kingscourt, Co. Cavan. Limestone is also widely used as a source of lime for the chemical industry.

The Four Courts Dublin, a fine example of a Wicklow granite building.

Part Three

Resources for Energy

Fossil Fuels - An Introduction

Millions of years ago, a number of substances were formed which have since remained buried in the earth, and which we now know as "fossil fuels". They are oil, natural gas, coal and turf. These are known as fossil fuels because they were formed from the remains of ancient plant and animal life. These substances are a vital resource to our way of life today, as they supply most of the energy on which we depend for heat, light, transport, and operation of machinery and appliances. Although the process which formed these fuels is still taking place, the replacement span is so long that we can consider these resources as non-renewable. At present rates of use, most of the world's known oil and gas stocks will be largely depleted by the year 2000 AD. Coal will last longer, but it has fewer uses than oil. Turf is not a great resource in world terms, and, like coal, has limited application. Oil, then, is seen as the life-blood of our civilization, and it is steadily ebbing away.

Governments are already anxiously looking for other sources of power. In Ireland, the Electricity Supply Board has plans for the building of a nuclear power station in Wexford. Large numbers of people have demonstrated against these plans and the government have yet to make a final decision. In Britain, scientists estimate that, by the year 2000, wave action could be generating power equal to the burning of 15 million tonnes of coal per year. The United States Government have set aside funds to investiage the possible uses of solar energy. These two energy sources are renewable and pollution-free. Research is taking place into ways of converting coal into a liquid fuel (synthetic oil), the use of alcohol-driven engines and the development of a practical type of electric vehicle.

Oil

OIL PRODUCTION

Oil was formed millions of years ago when large parts of the earth were covered in salt swamps. These areas supported lush vegetation and many small forms of animal life. Thousands of generations of plants and animals lived, died and sank into the mud. In time, the swamp was gradually covered over with layers of silt carried down from the hills in rivers and streams. The silt mixed with the plant and animal remains, and built up into layers many hundreds of metres thick; the lower layers were compressed and hardened by the great weight, and eventually formed into rock. This is what is known as *Sedimentary Rock.*

In places, vast lakes of swamp ooze collected between the rock layers, seeping through until they met a hard, impermeable surface. Millions of years passed, and the ooze was slowly changed into a thick black liquid, what we know as crude oil. Then dis-

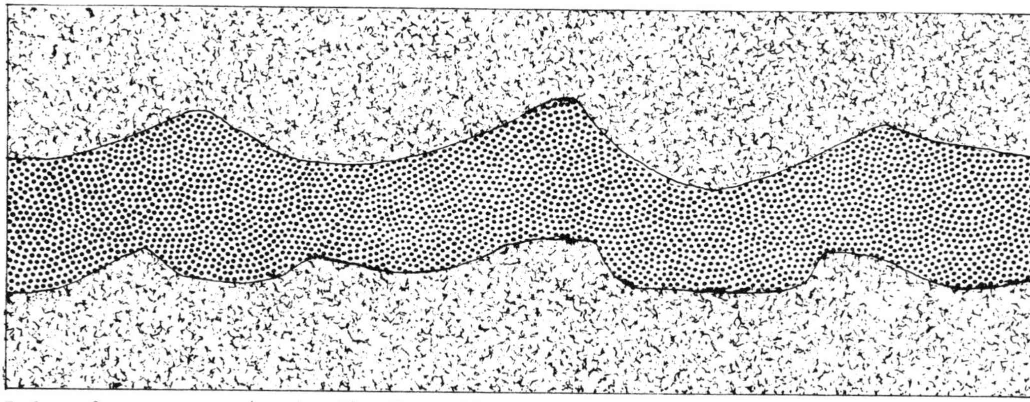

Lakes of swamp-ooze (crude-oil) collected between a layer of sedimentary rock.

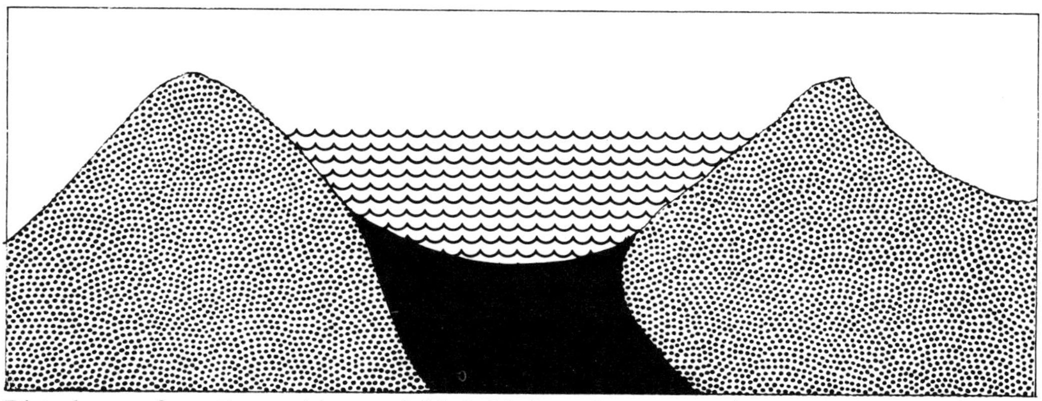

Disturbances from the earth's core folded this layer into mountains, the crude oil was trapped in the pockets called oil-wells.

turbances from the earth's core forced the layers of rock upwards, bending them into hills, plains and valleys. The oil was trapped in pockets between the twisted layers. These pockets are known as oil wells.

Sometimes, these underground pools come close enough to the surface for oil to seep up and form black puddles. Man has long used this crude oil for lighting. It was not until about a hundred years ago that scientists learned to drill wells and refine the oil into more useful products such as petrol and paraffin. Today, the main uses of oil are as a source of energy, it is used to generate electricity and as a lubricant for machinery and engines. When refined into petrol it is used as power for motor cars and other engines. With the help of scientists, the list of oil products increases all the time.

To meet the world's demand for oil, more and more crude oil has to be produced. Man has been drilling for oil for over a hundred years. Increased knowledge about rocks and rock formations have helped to choose the best places for drilling. Modern technology has enabled engineers to construct drilling and pumping equipment to recover oil from all locations: tropical swamps, the frozen wastes of Alaska and Siberia, and even from the sea-bed.

OIL CONSUMPTION

The world's consumption of oil has

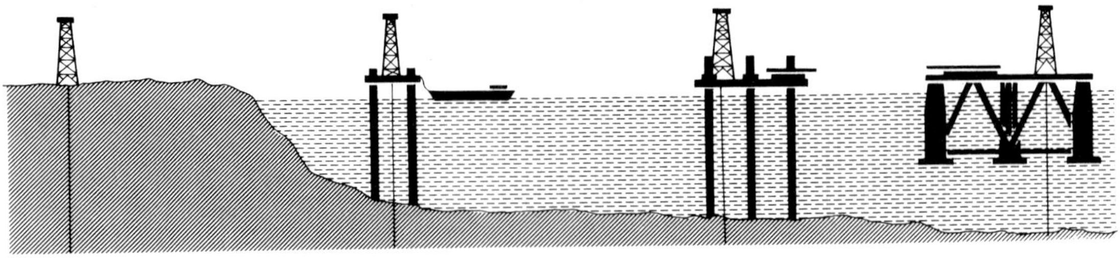

Four types of drilling rig: a land rig, an underwater platform with a drilling barge, a moveable platform, and a floating rig held by anchors.

increased dramatically in recent years:

In millions of tonnes:

1910	45m
1920	95m
1930	204m
1940	300m
1950	540m
1973	2835m
1978	3083m
1979	3120m

(Source: B.P. Statistical Review of the World Oil Industry 1979).

The main consuming countries are not, in general, the main producers. Western Europe, in particular, needs huge amounts of oil, but has almost no reserve of its own.

The Middle East, U.S.A., U.S.S.R. and Venezuela are the world's leading oil producers. In recent years, further large deposits of oil have been discovered in the North Sea, Western Siberia and Alaska. These oil fields have quickly been brought into production.

CONTROL OF THE WORLD'S OIL

Outside of the Socialist countries, control of the world's oil has rested on the major oil companies, often known as the "Seven Sisters".

"SEVEN SISTERS"

1. Esso
2. Royal Dutch Shell
3. Mobil
4. Gulf Oil
5. Chevron
6. B.P.
7. Texaco

For years, they largely determined the price to be paid to the producing country for each barrel of oil that was produced. They also decided the pace at which the oil would be extracted. Some of the oil producing states felt that they were not getting an adequate return for such a valuable resource. The oil states then formed an organization known as the Organization of Petroleum Exporting Countries. O.P.E.C. took joint action for the first time in 1970 and have become more powerful since. They decide the price which the oil companies must now pay for each barrel of oil. They also decide the pace at which the oil may be extracted. This means that the rich countries of the world can no longer get their oil at a cheap price.

The Rising Costs of Ireland's Oil Imports
in million £

1973 – £ 68
1974 – £210
1975 – £228
1976 – £298
1977 – £362
1978 – £347
1979 – £533
1980 – £800 (estimation)

(Source: Central Statistics Office)

The extra cost of oil imports has been passed on to the ordinary consumer. Within the last few years, the cost of petrol, heating oil and plastics have increased dramatically:

Retail price for Grade 1 petrol,
Ireland 1973-80*

1973 –	41.4 pence per gallon
1974 –	65.0 pence per gallon
1975 –	75.7 pence per gallon
1976 –	85.7 pence per gallon
1977 –	97.0 pence per gallon
1978 –	92.0 pence per gallon
1979 –	122.4 pence per gallon
1980 –	155.4 pence per gallon

(Source: National Prices Commission)

**Government Duty and V.A.T. account for roughly 50% of these prices. The amount of taxation on petrol changes with Government Budgets, but in this period, it has remained within a 40%-60% range.*

OIL SUPPLY AND DEMAND

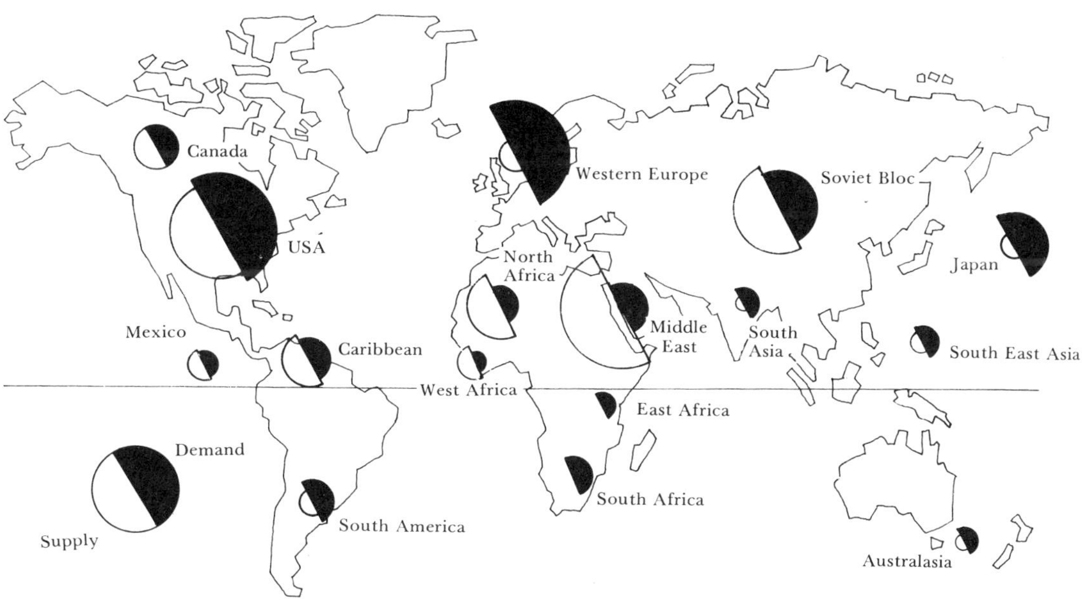

These increases in the cost of oil and its products have had a very bad effect on the economies of Western European countries, who import great quantities of oil. It has meant that industrial costs in general have risen, thus adding to the prices of all goods.

O.P.E.C.'s strong influence on the world's oil market also means that if anything happens to disrupt oil-production in these countries, the effects are felt all over the world. In 1980, the war between Iran and Iraq posed a grave threat to world oil supplies as the oilfields on the Persian Gulf were prime targets for military attacks. Governments in many non-oil producing countries, including Ireland, are making efforts to reduce their dependence on imported oil by encouraging conservation, use of alternate sources of energy and research into new sources.

Many of the O.P.E.C. countries use oil revenue to build hospitals, schools, factories and roads for the population. The standard of living for people in a country like Saudi Arabia has improved greatly. Some European firms are now under contract to Middle East oil-producing countries to build schools, hospitals, and harbours.

IRISH OIL

Drilling for oil in Irish waters began in 1970 in the Celtic Sea basin. In 1971, a significant Natural Gas deposit was struck off Kinsale Head; as oil and natural gas are usually found close together, there are high hopes that commercial quantities of oil will be struck. The most likely areas for this to happen will be the sea areas Rockall, Porcupine and the Celtic Sea. In the 1979 drilling season, two wells were drilled on the Porcupine Bank which produced oil. This well poses technological difficulties because of the depth of water and bad weather conditions. It would take a major find to make it worthwhile to develop equip-

IRELAND'S OFFSHORE OIL EXPLORATION

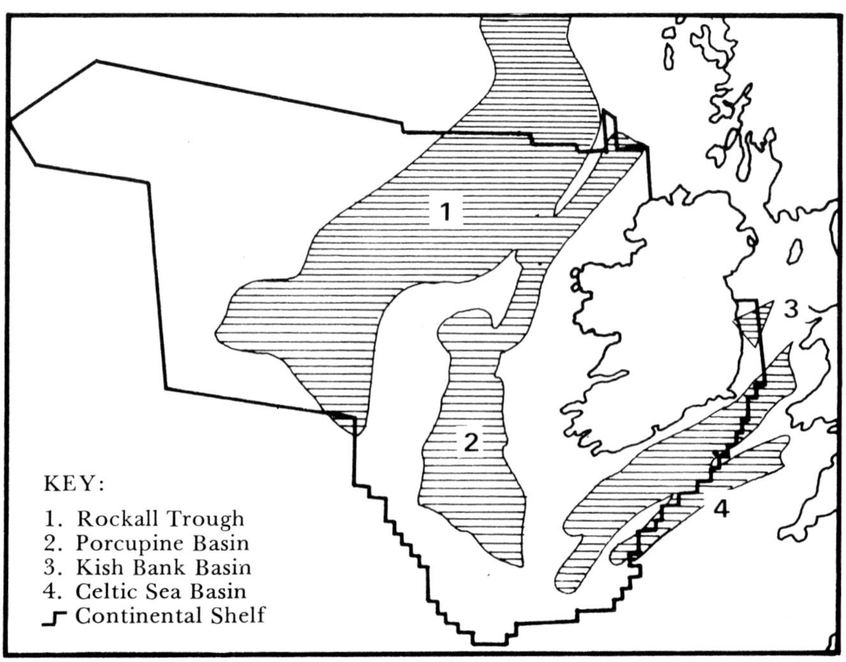

ment for drilling and pumping in such a difficult area.

The Marathon Oil Company (U.S.) holds the largest blocks of sea under oil prospecting licence. If oil is struck off the Irish coast, the use to which the resulting revenue will be put will be of major concern to everyone. The Norwegian Government takes up to 75% of all oil revenue in taxation, with the figure of 90% being mentioned as a future take. This income is then used to develop the whole economy.

Offshore oil exploration has become a major element in the search for mineral resources as Ireland has a high dependency on imported fuel for energy requirements.

REFINING

The crude oil that is extracted from a well must be refined before it is of any commercial use. In the refining process oil is broken down and changed into petrol, fuel oil for generating stations, aviation oil, lubricating oil and hundreds of other by-products. Oil refineries are generally located near great centres of population and industry, like the Rhine/Ruhr district of Federal Germany, Moscow industrial area, and the great conurbation of the N.E. United States. These areas have the greatest demand for oil products. Oil fields are generally located in remote areas, therefore the crude oil must be either piped for hundreds or even thousands of kilometres across land or transported by large oil tankers across the sea.

Ireland's Oil Refinery

The Republic of Ireland has a small oil refinery at Whitegate in Cork Harbour. This refinery was built in 1959 and is jointly controlled by Esso, B.P. and Texaco. The refinery supplies most of the Irish market with motor fuel, which means that Ireland is dependent on one refinery for most of its oil requirements. This dependence has lessened due to the E.S.B.'s decision to enter into contract for refined oil from the U.S.S.R. This oil is used in some of the generating stations in Ireland. As the country's demand for refined oil continues to rise, the building of another refinery has been proposed. Dublin Bay has been suggested as a possible site, because of the proximity of a very large population and industrial concentration. Many people objected to this because Dublin Bay is regarded as a Special Amenity Area and it was felt that the siting of an oil refinery would cause too many environmental hazards. Others argued that, if the refinery were properly controlled, it would provide much needed employment in the Port Area where unemployment is already very high. This refinery could be the basis for a large petro-chemical industry. Factories could then be built, using the oil from the refinery to make fertilizers, plastics, textiles, gramophone records, shoes, wallpaper, carpets and medical drugs. If another location could be found which would not pose a threat to the environment, an oil refinery could lead to industrial development and boost the economy.

TRANSPORTING OIL

Oil can be moved overland by piping, even over long distances, but, as the oil fields are generally very far from the countries where the demand is greatest, it is necessary to move huge amounts of oil overseas. This has given rise to the development of supertankers, the largest ships ever built. Some of them have up to

500,000 tonne capacity. With such large ships moving in the world's busy shipping lanes, the risk of sea collisions is increasing; a tanker can take several kilometres of sea space to stop, or turn.

WHAT HAPPENS AT THE REFINERY?

At the refinery, crude oil is heated in pipes inside a furnace, and the liquids that make up the crude oil turn into vapours at different temperatures. These vapours pass into a tall steel tower and as they rise slowly they cool down. As each vapour cools, it condenses (turns back into a liquid) at different heights in the tower, the liquids that boil at the lowest temperature coming off at the top. Each kind of oil is caught in a separate tray and removed by pipes leading out of the tower. This process of boiling and then cooling the oil again is called distillation.

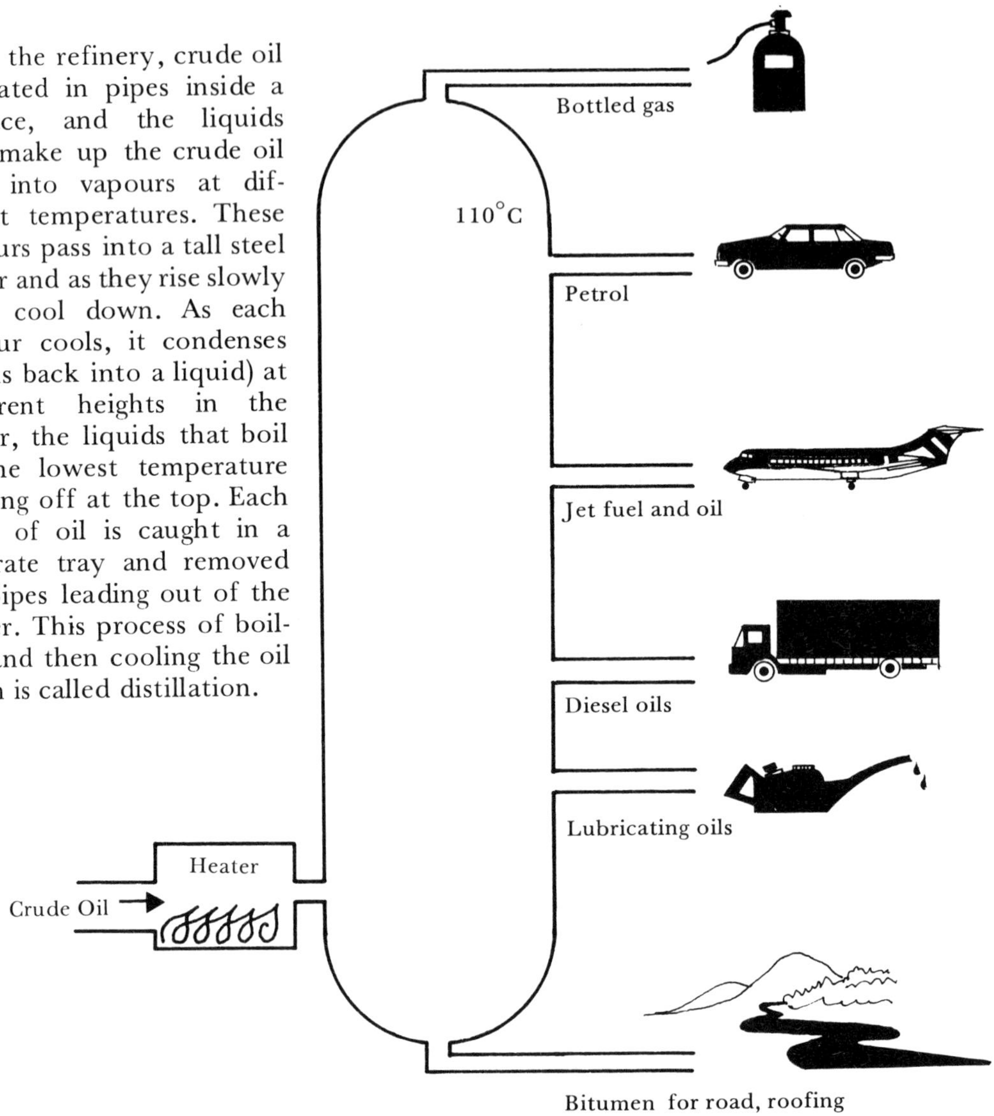

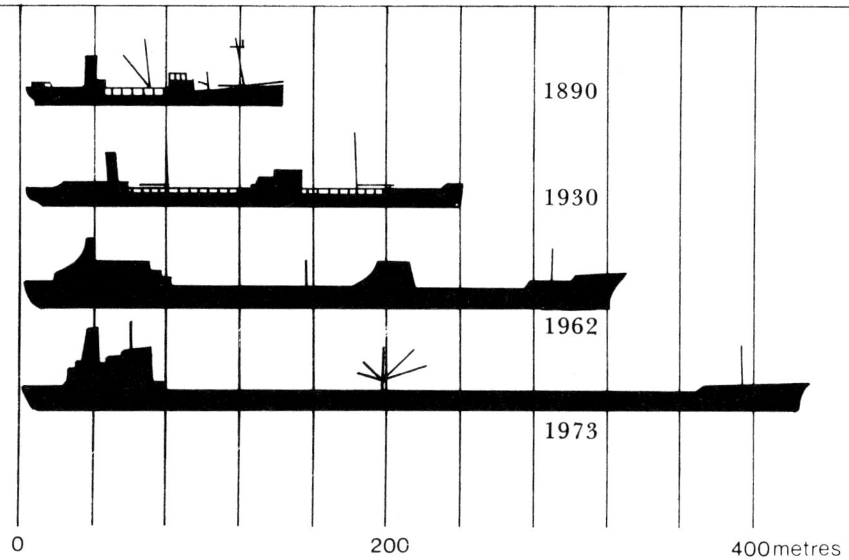

INCREASE IN THE SIZE OF OIL TANKERS OVER THE YEARS

A modern tanker is equipped with the most advanced equipment for navigation and control, so that it needs only a small crew. Living facilities for the crew members are very good indeed. The cargo is contained in dozens of separately sealed tanks, and many ships have special equipment for detecting leaks, fire, uneven or excess loadings and any other hazards. Despite this, accidents do happen.

The worst of all recent accidents involving a supertanker happened on the night of January 8th, 1979, at 1.00 am, when the tanker *Betelgeuse* exploded off Whiddy Island, in Cork, killing fifty people, destroying the jetty and causing a massive pollution problem. A public enquiry resulted in strong criticism of both the shipping company (Total) and the terminal operators (Gulf). Oil is a dangerous commodity, it requires careful management.

POLLUTION FROM OIL

Oil is one of the newest and most serious chemical sources of pollution of the environment, especially the sea. When oil is spilled from a tanker in the open sea, it floats on the surface and can travel hundreds of kilometres from the point of origin. This oil can destroy animal and plant life; when it is swept ashore, it covers the beaches with a thick black slime. Detergents which are used to break up the oil slicks can also be harmful to the environment. A serious oil spillage can ruin fishing in an area thus threatening the livelihood of fishermen. Recently attempts have been made to make the oil companies responsible for damage and liable for paying out compensation if accidents occur. As long as oil is carried by tankers, there is always a danger that spillages and explosions will cause great pollution. In 1978, two major oil spillages occurred. The coasts of Brittany in Northern France and East Anglia in England were badly polluted by crude oil discharged from tankers separately involved in sea incidents. The resulting damage severely affected fishing grounds and beaches in both countries. The total

Tanker disaster leaves 50 dead at Whiddy Island

From Dick Hogan, in Bantry

THE DEATH TOLL in the Bantry oil tanker disaster stood at 50 last night, after an intensive search failed to find any survivors of the massive explosion which engulfed the French vessel Betelgeuse early yesterday morning. The entire French crew of 43 and seven local Gulf Oil workers died in the country's worst fire disaster this century.

Previous disasters

Ma[illegible]h, 1974 — 33 people [illegible] following bomb explo[illegible] in Dublin and Monag[illegible].

Mar[illegible] 24th, 1968 — 61 people [illegible] when a Viscount 803 [illegible]ashed into the sea [illegible] off Tuskar Rock.

Septe[illegible] 10th, 1961 — 83 people died when a Presiden[illegible] DC 6 crashed [illegible] Shann[illegible] Airport.

Febru[illegible]st, 1953 — 128 [illegible]ople died when the Prin[illegible] Vict[illegible] sank in heavy [illegible] at the entrance to Bel[illegible] Lough.

[illegible]bruar[illegible] 1943 — [illegible] died [illegible] fire at [illegible] Poor Clare [illegible] Co. Cavan.

[illegible]ember [illegible] 1926 — [illegible] died in a fire at [illegible] in Drumcollogher, [illegible]rick.

There was no clear indication last night as to the cause of the accident but a Gulf Oil spokesman confirmed that the explosions followed a fire on board the tanker. The fire broke out after 80,000 tons of crude oil, two thirds of its cargo, had been unloaded into the Whiddy Island terminal.

The president of the tanker company Total, Mr Louis Bouzol, confirmed last night that the Betelgeuse did not have the safety "inert gas" system which is now used in tankers to prevent explosions of the combustible gas which accumulates as oil is being discharged. This safety procedure was not standard when the tanker was built in 1968, he said.

Despite the devastation to the tanker there appeared to be little pollution in the bay last night. The fire in the middle of the tanker appeared to be burning off the escaping oil.

When the intensive search of the bay was suspended at darkness last night 18 bodies had been recovered and flown by helicopter to the Regional Hospital in Cork. The victims were so badly burned that only two had been identified last night. One of those identified was the wife of the ship's cook and the other was identified through a passport found on the body.

Inquiry promised

[illegible] Pat Joy said he noticed a fire on board the tanker at 12.30, and then raised the alarm through his superiors and brought a well-planned emergency operation into action.

Gulf said that at 12.55, control tower staff on Whiddy Island heard a "crack" coming from the mid-section of the tanker. Within the space of three minutes, this was followed by an explosion, followed again two or three minutes later by a massive flash flame. Oil began to seep from the tanker and control tower staff pressed an alarm button activating a siren and foam extinguishers.

Garda Joy told reporters that at about 12.30 on Sunday night he was at the pier in Bantry investigating a drowning accident when he saw what he believed was a fire on board the tanker.

cost of the spillage, affecting as it does the livelihood of thousands of people, ran to millions of pounds.

Ireland, too, has suffered from the effects of oil pollution caused by tanker spillage. During the night of October 21st, 1974, there was a leakage of oil from the tanker *Universe Leader*, which was loading a cargo of crude oil at the Gulf Oil terminal off Whiddy Island. About 2,500,000 litres had been spilled probably when a valve was left open by mistake. This oil caused severe damage to the area around Bantry Bay, which is one of Ireland's most beautiful places. Fishing grounds were ruined, thousands of birds were covered with oil and many of them died, and the shorelines were covered with thick crude oil.

Gas

Most types of gas used in homes and industries are manufactured, artificial gases. They are produced from other basic resources such as coal or crude oil. Natural gas is, however, a resource in itself, being a fossil fuel extracted more or less directly from the earth. This process involves drilling and recovery in much the same way as oil, indeed oil and natural gas deposits are often found together. Of the total known reserves of natural gas, about one-third are located in the U.S.S.R. and a quarter in the Middle East. The U.S. is the world's largest gas market and consumes more than half of all the gas produced. Nevertheless it has only small reserves of its own and even these are running out. Western Europe has 10% of the reserves and uses about 15%.

Gas reserves are often located far from the areas where the demand exists. Therefore gas needs to be transported long distances. It can, of course, be piped, North sea gas is piped ashore to Scotland and Norway. But when the distances involved are very great, the gas is first liquified, then transported, and finally changed back into gas again. The Kinsale Head gas field off Co. Cork is Ireland's only workable gas discovery so far. This produces gas which is piped ashore and distributed by the Cork Gas Company for domestic uses—cooking, heating etc. Some of it is used to manufacture fertilisers in a factory specially set up for the purpose, and the ESB is building a power station to make use of the gas. In December 1980, it was announced that the gas field was 40% bigger than originally predicted. A natural gas line to Dublin, Limerick, Belfast and other major urban areas could be developed. Discoveries of this kind help our economy and reduce our dependence on foreign sources of energy.

WORLD CONSUMPTION OF GAS

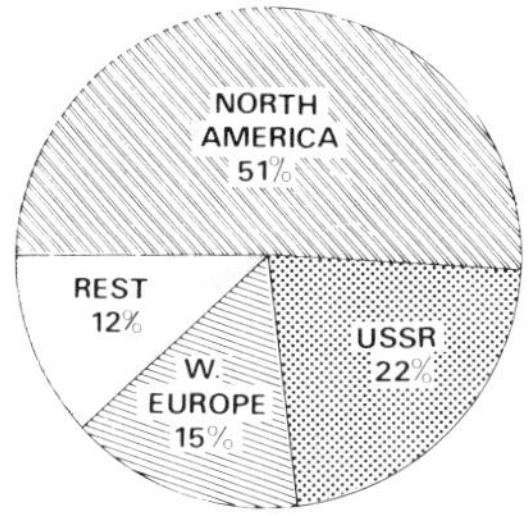

ESTIMATED WORLD GAS RESERVES

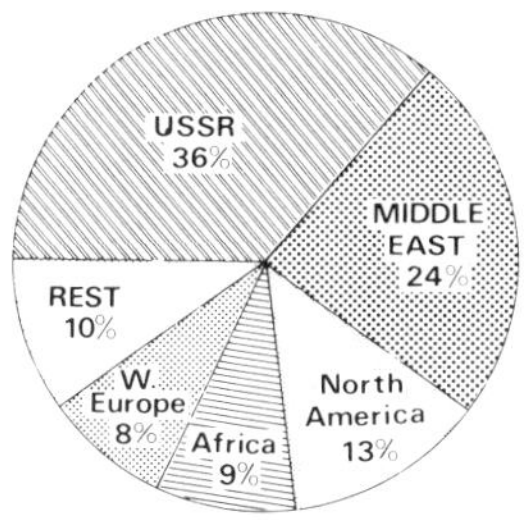

GAS RESERVES IN WESTERN EUROPE

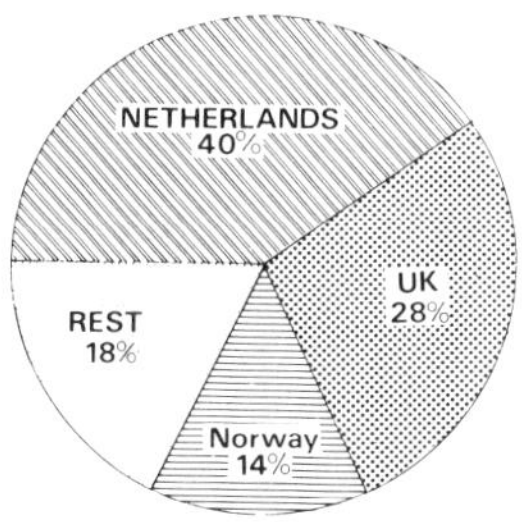

Coal

Coal was formed in a similar way to oil, and at approximately the same

Inside an old coal mine at Ballingarry, Ireland.

time in the earth's history. It lies under the ground in long, thick seams, sandwiched between layers of slate or shale. It is usually mined from undergound shafts, but occasionally from open-cast pits where the coal lies near the surface.

Coal was once the industrial world's chief source of energy. It was later replaced by oil, as liquid fuel was cheaper, more convenient, cleaner and had more varied uses. The growing fear that oil supplies will soon begin to dwindle has caused scientists and engineers to look again at coal as a resource.

The world's coal resources are vast and over 90% of them are believed to be located in the U.S.S.R., the U.S.A. and China. They are probably equal to 20-50 times the amount of recoverable oil.

We are all familiar with the use of coal in lump form in our fireplaces and in furnaces. However the major users such as electricity generating plants and cement industries burn coal in powdered form. The fine dust is blown into the burners in a jet of air. Coal has long been used as a raw material to make gas. Nowadays, research is taking place into ways of converting coal into a liquid fuel, a synthetic oil. Processes for doing this already exist, but they are either wasteful, expensive or unreliable. If an efficient way is found of making this conversion, coal could become, once again, the world's main source of energy.

Unfortunately, the use of coal has many environmental drawbacks. Coal mining of any kind causes landscape damage and disturbance, open cast mining cuts huge slices off the surface, underground methods leave great heaps of waste material. The mining companies should, of course, landscape the affected areas, repair any damage done to drainage systems, plant trees, but in the past this has seldom been done. The great slag heaps of waste material which are often allowed to accumulate near coal mines are also a source of danger. In October 1966, in the Welsh village

WORLD COAL RESERVES *(Billions of Tonnes)*

	Hard Coal (1)	Soft Coal (2)
U.S.S.R.	166	107
North America	128	60
China	101	-
India	11	1
W. Europe	37	11
E. Europe	4	27
Africa	12	-
S. America	-	1
Japan	1	-
Oceania	14	11
Rest of World	2	1
Total World	476	219

Source: World Energy Conference Survey of Energy Resources, 1974

(1) Includes anthracite and bituminous coals

(2) Includes sub-bituminous, lignite and brown coals

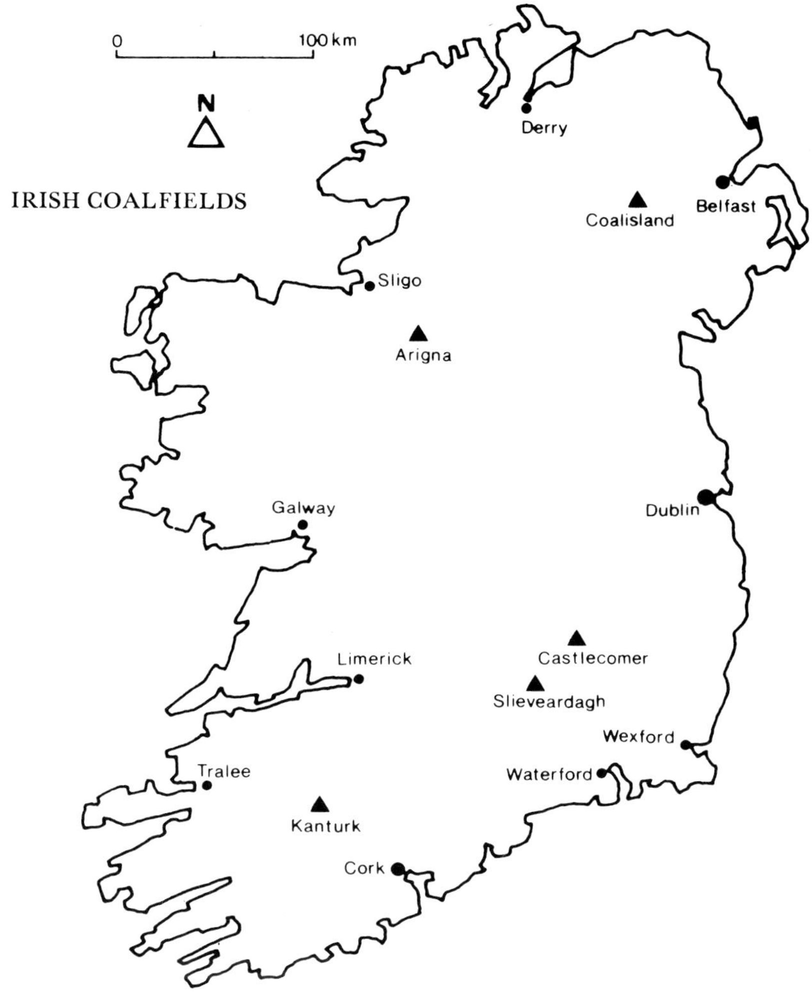

of Aberfan, one of these heaps collapsed, completely covering a nearby school and killing many of the children there. Coal also causes air pollution, when burned it emits sulphur dioxide and other gases. This problem can be overcome by use of special flues and by processing the coal before burning to make "smokeless" fuel.

Ireland has some coal resources but they are unfortunately very variable in quality and occur in thin, scattered seams. This means that most of them are not economical to recover and use. The Leinster coalfields have high-quality anthracite coal. The main colliery in this area is the Ballingarry mine, which first opened in the last century, and began operations again during the Second World War. It closed again recently. An open-cast mine in Castlecomer, Co. Kilkenny, produces 30,000 tonnes annually, and Rossmore colliery produces 20,000. The country's main coal mine is at Arigna, in Co. Roscommon. This produces 60,000 tonnes per year: it is used for generating electricity in a nearby power station. This coalfield has large reserves, possibly 8,000,000 tonnes of a very low-grade material. It was considered useless, but the ESB has recently invented a new process for burning this coal, and it should now prove a valuable resource.

Peat

Turf, or peat, is Ireland's other known native "fossil fuel". This resource is taken from the boglands which cover about one-seventh of the surface of the country. Peat is not as old as oil or coal; in fact, it probably represents a stage in the process of coal-making. There are two kinds of peat-bog in Ireland. The vast "raised bogs" of the central lowlands have an average depth of up to seven metres of turf. Much of our bogland, however, is the thin "blanket bog" found in the west and in higher mountain areas, with an average turf depth of only 2.5 metres. The Soviet Union has 60% of the world's peat resources and yet Ireland, with only 1% of the total deposits is the second largest turf producer in the world. Irish expertise, equipment and machinery are currently being exported to other countries who want to develop their peat resources. Our lead in this field is due to our long tradition of turf cutting in Ireland, where a wet climate and lack of coal and oil deposits resulted in the development of a local fuel.

For hundreds of years, turf was cut and saved in the bogs and used for cooking and heating in homes throughout the country. Even today, a farmhouse with its own family-harvested turf stack is a common sight, particularly in the west of Ireland. The great development of our peat deposits came with the creation of Bord na Mona. Since 1946, this state-sponsored body has been acquiring all the large raised bogs for mechanised harvesting. Giant machines which were designed by Bord na Mona have been put to use in the bogs, cutting great quantities of turf. At present, 900,000 tonnes of sod turf and 3,150,000 tonnes of milled peat are produced annually. Milling is a process by which the sur-

An electricity power station fuelled by peat won from the nearby bogs, Ferbane, Co. Offaly.

face of the bog is ground into a fine dust; this dust dries very quickly. Milled peat can then be compressed into blocks for burning, the convenient briquettes which are widely used for domestic fires today. Most of the milled peat produced is used to fire the ESB turf-burning power stations. Here it is burned in furnaces to produce steam, to turn turbines, so producing electricity. The seven largest turf-burning stations are all close to their sources of fuel. Those at Portarlington and Allenwood use sod turf; Ferbane, Rhode, Shannonbridge and Bellacorick burn milled peat. Lanesborough uses both. There are several other small stations in the west using hand-harvested turf. Bord na Mona directly employs about 4,500 people. Many other jobs in power generation, transport and marketing depend indirectly on the peat resource for their existence.

At present, 70,000 hectares of bogland are in production or under development. Beyond that, only small areas are suitable for mechanised harvesting. As with all fossil fuels, peat is nonrenewable. By the year 2000, most will be gone, the former bogs being reclaimed for use as new farming land.

Nuclear Energy

There are now about 200 nuclear reactors generating electricity in twenty countries throughout the world. Hundreds more are planned or being built. Nuclear power was first used in 1945, when the Americans dropped atomic bombs on the cities of Hiroshima and Nagasaki in Japan, causing horrific damage and casualties. This horrific event of World War Two led to the development of nuclear power as an energy source. Nuclear power is best suited to generating electricity.

The basic job of a nuclear power station is similar to that of a coal, turf or oil-burning one. A fuel is used to produce heat, this boils water to make steam, which turns turbines.

At one of the anti-nuclear protests staged at Carnsore Point, Co. Wexford.

The difference is in the fuel used. In this case, the fuel is an element called *uranium*. This substance is the basic resource that supplies nuclear energy. Uranium is not burned, like coal or oil. In a nuclear power station, heat comes from a chain reactor in which uranium breaks apart, thus releasing heat; this breaking up is called *fission*. It happens because uranium is an *unstable* element. If all the energy in one kilogram of uranium could be released, it would be equal to the energy produced by burning 3,000,000 kilograms of coal. In a reactor, rods containing uranium are placed in a specially sealed area, the *core* of the reactor. Here the rate at which fission takes place can be controlled by the amount of the rods exposed. Eventually, all the uranium is used up, and the rods must be replaced. This happens every one to five years. The process of fission creates new substances which were not there in the original fuel. They are highly radioactive and very dangerous. A complicated cooling system has to be used to ensure that the heat generated in the core is controlled at all times.

Nuclear reactors can supply huge amounts of energy from very small amounts of raw material. They seem to offer a very good way of conserving other resources such as coal and oil. The operations of nuclear reactors, however, have led to protests from environmental groups.

Several difficulties have arisen in the operation of nuclear power stations over the years. Getting rid of the used uranium rods and other wastes such as radioactive cooling water, has been a major problem. A radioactive substance does not harm anyone by touch, it emits harmful rays, in the same way as the sun emits heat. At present, these wastes are mostly buried in old mines, or dumped in deep parts of the sea in sealed containers. Some of them are "cleaned" at special plants. Disposing nuclear wastes is very costly. A nuclear power station cannot explode like a bomb. The worst accident that can happen is called a core melt-down. This means that if the cooling systems fail, the resulting heat can melt the uranium rods and their casings, then burn through the reactor's outer walls and burn into the earth below. This would release clouds of highly radioactive steam and gases into the atmosphere. As yet, a melt-down has never actually occurred, although there have been several "near accidents".

There have been many other problems in the use of nuclear reactors; costs of waste disposal are very high and many reactors are inactive for long periods, needing costly maintenance. The greatest problem of all is what to do with a reactor when it is worn out. After about 25 years of use, a reactor is too radioactive for safety. So far, nobody knows what to do with such a station to make it harmless. At present, a disused reactor is simply filled full of concrete, walled up and left under guard. All in all, nuclear energy has not proved the "almost free' source of power it was expected to be.

Ireland, at present, has no nuclear power station. The ESB has plans to build one at Carnsore Point, Co. Wexford. This plan has led to great controversy. Many people are afraid of possible effects it might have on the environment, others fear that waste materials may not be properly disposed of and endanger the people living nearby. On the other hand, everybody wants cheap and reliable energy, but is it worth the risk?

Energy for Free

There are three sources of energy open to us which are infinitely renewable, they are wind, wave and solar power.

Wind: Wind power may be simply harnessed by the use of a windmill. This is a tower with a propeller or sails mounted on a spindle. The wind turns a propeller, which turns a generator, and so produces electricity. Windmills can, of course, be used for other purposes, such as milling grain and pumping water, as in the Netherlands.

Ireland is one of the most windswept countries in the world, particularly on the west coast. Experimental windmills are already in use, and are showing promising results.

Wave: The west coast of Ireland has more potential for harnessing wave power than many other parts of the world. It is estimated that approximately 40 kilometres of wave generators could supply as much electricity as the total output of the ESB at present. This has advantages over wind and solar power, in that the tides come and go regularly. Research into this area is at an early stage, but the signs are encouraging.

Solar: As well as being used indirectly, solar energy can be used directly. Appliances for collecting and using the heat of the sun have long been used in the hot countries of the world. This heat is an ideal resource if it can be used properly: each year the sun sends to earth 100,000 times the total amount of energy used by the world's population.

A solar collector is basically a black metal plate in a shallow glass box. Water runs through tubes attached to the plate. The sun heats the plate, and therefore the water in the tubes is heated and can be circulated to radiators and other heating outlets. So far, solar collections are suitable for small scale uses only, such as heating homes and other buildings. A collector large enough to generate 100 mw of electricity would have to cover an area of about 2½ sq. kilometres! Ireland is well suited to developing solar energy use. Japan, which is not a hot country, has 3½ million solar collectors in use. We may yet see solar panels in widespread use on the roofs of houses in Ireland.

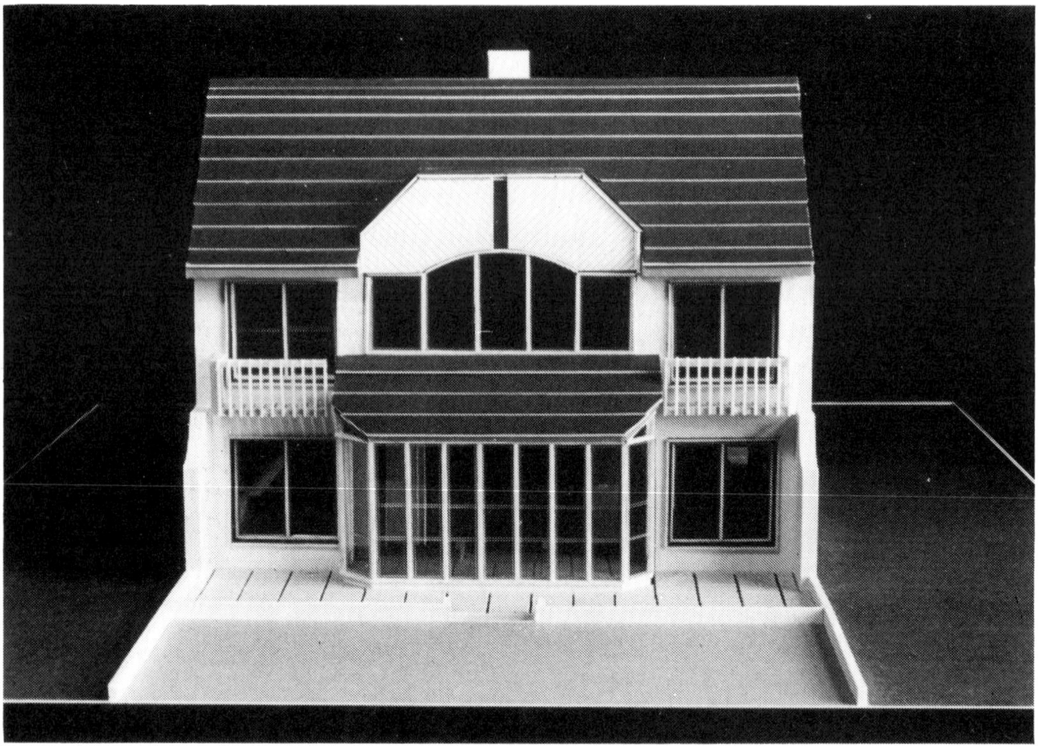

An alternative to solar collectors; the windows and glass conservatories of this house face south to capture the heat of the sun, heavy internal walls and chimney breasts distribute the heat to the various rooms.

Part Four

Resources in the Wicklow Area
~ A Case Study

PRELUDE

Still south I went and west and south
again,
Through Wicklow from the morning
till the night,
And far from cities, and the sights of
men,
Lived with the sunshine, and the
moon's delight.

I knew the stars, the flowers, and the
birds,
The grey and wintry sides of many
glens,
And did but half remember human
words,
In converse with the mountains,
moors and fens.

J.M. SYNGE (1871-1909)
Poems and Translations

Wicklow - An Introduction

Ireland is shaped like a saucer, with a low centre and a high rim of mountains. This rim is almost continuous along the West coast, with the mountains of Kerry, Galway, Mayo and Donegal including most of the high peaks in the country. Along the East coast there are gaps in the rim, as in Co. Dublin. These eastern lowlands have provided a route for Celts, Vikings, Normans and English invaders. Trade with Britain and Europe has long been focused through the eastern ports. As a result, this region has become prosperous and well-developed. Parts of the East coast are high. The main blocks of mountain are the Mourne Mountains of Co. Down and the Wicklow Mountains.

Co. Wicklow, to the South of Dublin city and county, has the most varied landscape in Ireland. In the East it is lined with sand and gravel beaches along the coast of the Irish Sea; in the West it joins the Central Plain of Ireland. It has rivers, lakes and marshland; neat farmland and bleak mountains; native Irish deciduous forest and newly-planted coniferous trees.

COUNTY WICKLOW RESOURCE INDUSTRIES

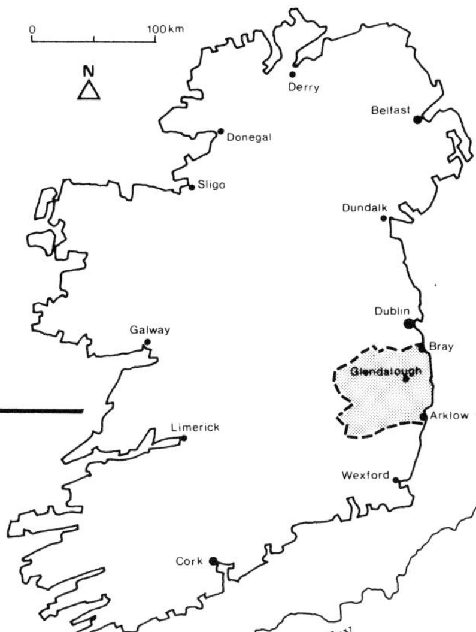

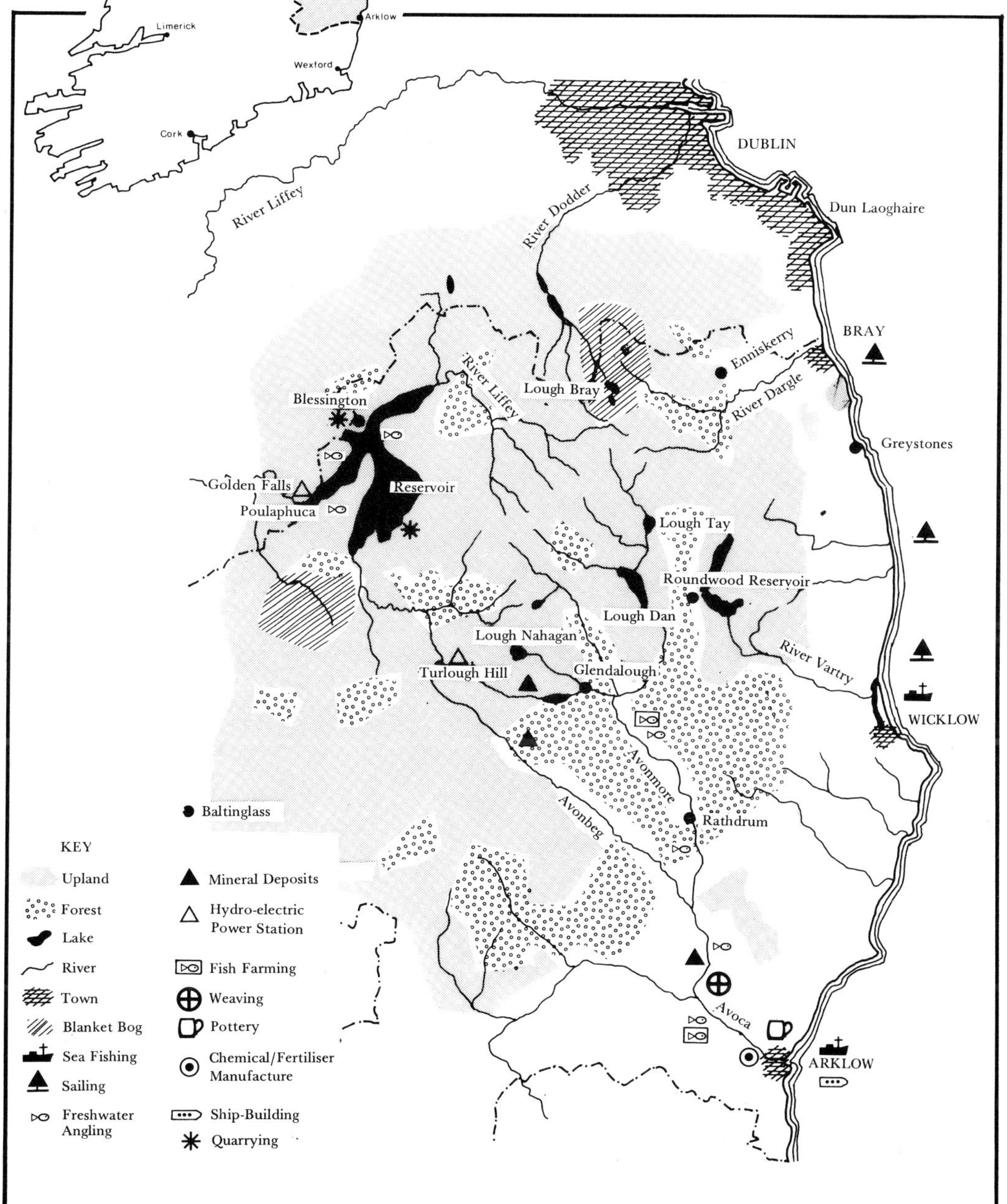

The way of life in Wicklow reflects this same variety. Some Wicklow people travel to Dublin to work in factories and offices. Some are farmers, rearing cattle and growing crops on the lowlands, or tending herds of sheep on the mountains. Some farms are small and worked by hand; others are large and mechanised. Other Wicklow people earn their living in fishing, in forestry, in tourism and in craft work, as well as in factories, shops and offices. If Dublin and Wicklow are compared, the differences between the two areas are immediately evident. Dublin is mainly lowland, Wicklow is mainly highland; Dublin is mainly urban, Wicklow is rural. Dublin has a relatively high density of population, Wicklow contains large unpopulated areas. On closer inspection, it becomes clear that Dublin and Wicklow have much in common, that they share many features, and that they depend on each other to a very great extent.

The rivers and lakes of Wicklow provide Dublin's water supply and part of its hydro-electricity needs. The farms of Wicklow provide an important part of Dublin's fresh perishable food such as milk, fresh vegetables and eggs. Wicklow turf is burned in some Dublin homes. Wicklow trees provide some of the timber and much of the chipboard used in housebuilding in the city. Wicklow's greatest asset is its natural suitability to tourism: mountains, lakes, forests, beaches and the sea. For most Dubliners, Wicklow is a great natural playground on the city's doorstep.

In return, Dublin provides a market for the farmers and fishermen of Wicklow. The visiting Dubliner helps to create employment in shops, restaurants, hotels and other services. Wicklow people work in the city, city workers commute to new homes in North Wicklow. The economies of the two counties are dependent on each other.

HIGHLAND–LOWLAND

Wicklow is not just a large mountain range. The county has a great variety of landscapes, and a matching variety of lifestyles for its people. Basically, the county can be divided into highland and lowland areas. Lowland Wicklow is located in two distinct regions. To the east is a lowlying plain between the mountains and the sea. It is 50 kilometres long from North to South and 8 kilometres wide. There are many large towns in this part of the county, communications such as roads and railways are extensive and the sea has had a major influence on the way of life. The towns of East Wicklow have many industries, large new housing estates for Dublin commuters and a growing range of shops, stores and entertainments. Bray is one of the

Modern housing at Newtownmountkennedy in the heart of scenic Wicklow, many of the inhabitants commute to Dublin to work.

most rapidly developing towns in Ireland.

To the west of the mountains is a landscape of low rolling hills which gradually merge with the Curragh of Kildare and the Central Plain of Ireland. It is an inland area and suffers from frost and snow in winter, but soils are good enough for sheep fattening and dairy farming, and some wheat growing. Many of the villages in West Wicklow have regular fairs and markets and are large enough to be called small towns. Blessington, Baltinglass, Kiltegan, Dunlavin, Shillelagh and Donard are all connected by a network of good roads. A number of these settlements were originally established as estate villages by the landlords who owned large farms and estates in West Wicklow. The villages are very neatly laid out with straight streets, central squares and market areas. Blessington is a fine example of a West Wicklow town with a long straight tree-lined main street, a market square with weighbridge and market house and buildings of Wicklow granite, which have a continuous and pleasing roofline. In the market square a monument stands to the landlord who established the town. Blessington, like many other villages in the area, dates from the Georgian period. West Wicklow is now being inhabited by more and more Dubliners who are attracted to the beauty and amenities of the area and who are prepared to drive up to 40 kilometres to work each day.

Highland Wicklow occupies a large proportion of the county. Must of the land is over 250 metres high and is therefore less suitable for human habitation, the growing of crops or the keeping of cattle. Communications are poor and many areas are often cut off by snow drifts in winter. There are no towns in the area, and most of the small population live in small isolated villages. The area has a very high annual rainfall and soils are poor. Many of the gentler slopes have been forested. Sheep freely roam the bleak windswept high ground where only heather and coarse grass can grow. Rocky outcrops and steep slopes make farming impossible in some parts. The villages have no industry and provide only basic services such as shop, pub, church and school. Despite all of these drawbacks, the highland area contains most of the natural resources of the county; some of these, as we shall see, are vital to the continued growth of Dublin city.

THE LANDSCAPE

Wicklow was originally an area of sedimentary rocks such as sandstone and shale. Millions of years ago, molten material from deep under the crust began to force its way upwards, and flowed along the layers of the sandstone. It formed a huge mass under the surface. The sandstone and other rocks through which it flowed were baked and compressed to such an extent that they changed completely into new rocks, *Quartz* and *Mica-schist*. The molten material slowly cooled, and hardened into the rock we call *Granite*. Over millions of years since, wind and rain, frost and snow, rivers and ice have worn away the outside layers and exposed the domes of granite. Parts of the old covering survive in places, such as the pointed quartz peak of the Sugar Loaf, and the mica-schist layers that flank Djouce and Maulin Mountains. Since the formation of the rocks, the landscape of Co. Wicklow has been influenced by the weather and the effects of the Ice Age. Weathering has worn the hills into their familiar dome shapes, and contributed to the formation of their covering of blanket bog with its heather, fern and fraughan vegetation. The effects of weathering can be clearly seen today on many of the mountain tops. Here, the peat covering is being stripped away, exposing the granite underneath and giving little islands of turf called "Peat Hags".

During the glacial period, Wicklow was covered by a layer of ice. As it melted and the glaciers retreated, deep valleys such as Glenmalure and Glendalough were carved out. In Glencree, a glacier moved down the valley, leaving great hollows that formed lakes called corries, for example, Lough Bray Upper and

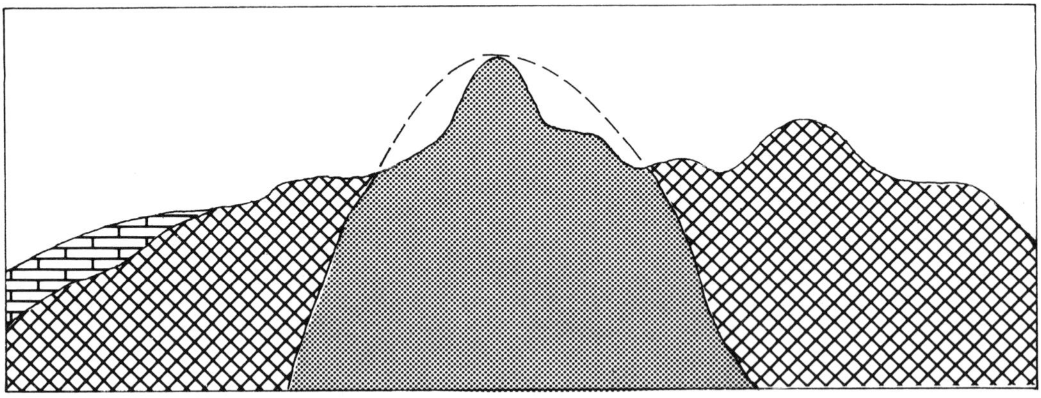

The dotted line represents the original rock, quartz, mica-schist, which was weathered down to expose the granite.

Granite.
Slates. Quartz.
Limestone
Sedimentary.

Peat hags and exposed granite bedrock at the summit of Tondruff North, at Glencree.

Lough Bray Lower. As the ice melted, heaps of boulders, stone, gravel and sand were left behind. These are called moraines. There are moraines between the two lakes, Upper and Lower Lough Bray, and in several other places along the sides of Glencree.

Resources in Wicklow

County Wicklow, with its varied landscape, has a wide range of natural resources. Some of these resources benefit the people of Wicklow and help them to earn a living, others benefit the people of Dublin and the country as a whole. Some resources are so close to depletion that they must be carefully conserved, others have been exhausted and must now be renewed. The resources of Wicklow include:

- *Water*
- *Land*
- *Metals and Other Minerals*
- *Wood*
- *Stone*
- *Turf*
- *A clean environment and amenity value.*

Most of these assets are located in the highland areas, the people and factories who use them are located in the lowlands, in Dublin, or far away from Wicklow.

WATER

Water Supply

Many of the large cities of the world have great difficulty in maintaining a water supply for industrial and domestic uses. Dublin is fortunate in that it has an assured water supply from Co. Wicklow for the foreseeable future. *Precipitation* (Rain, snow, sleet, hail) is very high in the mountainous Wicklow area. The annual rainfall in the low land of North Co. Dublin is only 915 mm, but in parts of Wicklow it is almost twice that figure, up to 1500 mm in places. Wicklow provides an ideal water catchment area because it consists of natural rock, soil and slopes. Water can soak into the ground to collect underground and reappear as springs, or it can run off on the surface as streams which may combine to form rivers. A built-up area does not allow this natural gathering and collecting of water to take place.

All of Dublin's water supply originates in three Wicklow rivers:

The River Vartry was dammed at Roundwood over 100 years ago, creating a large reservoir. Seventy-five million litres of water are supplied to

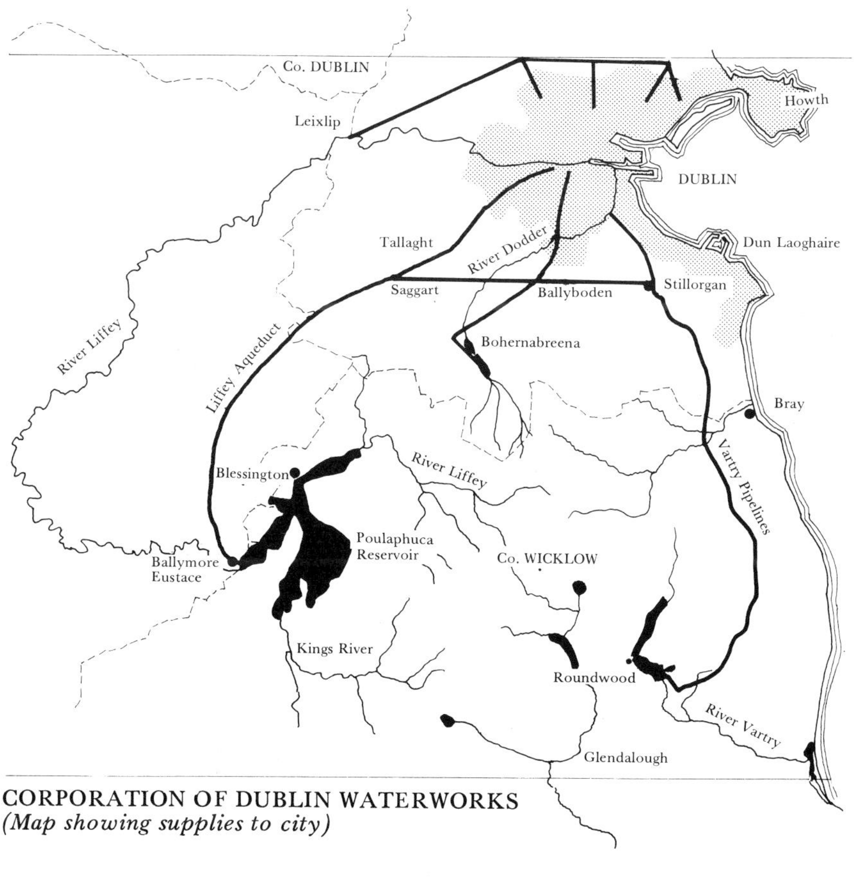

CORPORATION OF DUBLIN WATERWORKS
(Map showing supplies to city)

This photograph shows the giant scale of the engineering project that created a power station at Turlough Hill. The artificial reservoir on the right is used to store water; at peak periods it is allowed to fall through underground tunnels to the lake on the left. On the way it turns turbines to produce up to 1500 megawatt hours of electricity.

the city each day. The water is purified and filtered and piped to Stillorgan, from where it is distributed.

The River Liffey is dammed at Ballymore Eustace, forming a huge reservoir known as the Blessington Lakes. The reservoir was created in 1936 by flooding the valley. The local farmers were resettled elsewhere. In a dry summer, the outlines of fields and houses may be seen when the level of the lake falls. One hundred million litres of water are piped daily from this reservoir to Dublin.

The River Dodder rises on Kippure Mountain in Co. Wicklow, and is dammed at Bohernabreena near Tallaght to form a reservoir. The water is purified nearby at Ballyboden and 22 million litres are supplied to the city each day.

Water Power

Most of the Wicklow Rivers are not large enough to generate hydro-electricity in large amounts. Two power stations have been built on the River Liffey. One of these is in County Dublin at Golden Falls, the other is at Poulaphouca and shares the water supply from the Blessington Reservoir. These are standard hydro-electric stations.

A new type of power station was opened at Turlough Hill in 1973. An artificial lake was built on top of the hill and a system of tunnels leads down inside the hill to a power station at the bottom and finally to a glacial lake, Lough Nahanagan, which acts as a water storage reservoir. When demand for electricity is at a peak in

Dublin, usually during the late afternoon, millions of litres of water are released from the artificial lake to cascade down through the tunnels turning large turbines to generate electricity. The water is pumped back up from the storage reservoir at night. This pumping, of course, consumes a huge quantity of electricity, but demand is much lower at night and the electricity is therefore 'cheaper'.

Fishing

Sea Fishing was once the major industry along the Wicklow coast. Fleets of trawlers from Arklow and Wicklow ports ranged around the coast of Ireland, and Arklow sailing drifters caught mackerel and herring

Trout farm, Woodenbridge, Co. Wicklow.

as far afield as Scotland and the Isle of Man. Times have changed. The harbours at Wicklow and Arklow are constantly silting up and are not able to accommodate the large modern trawlers such as those based in Dun Laoghaire and Howth. Four hundred craft fished the Wicklow coast at the turn of the century. There are now only nine trawlers, all working out of Arklow. Although it no longer ranks as a major port, Arklow still looks very much a typical fishing town, though its inhabitants now earn their livings in factories. Industries associated with the sea still survive. Boatbuilding is a thriving industry at Arklow and young men from the area still sign on with coasters and cargo ships working out of Irish ports.

Inland Fisheries

The rivers and lakes of Wicklow are unsuitable for commercial freshwater fishing, but they offer the angler good sport in beautiful surroundings. They are thus a valuable resource, as they draw visitors to the area. Trout, and coarse fish such as pike and perch are the main species caught. Trout farms are operated at Woodenbridge and Knockrath. In these farms, fish are reared under controlled conditions until they are large enough for fishing. They are then released into the rivers, or sold for the table. This is a valuable industry, as demand for trout in the shops is very high.

FARMING

The land of Wicklow remains the county's single most important resource. Many factors influence the kind of farming carried out in various parts of Wicklow. Soil type, slope, aspect, distance from Dublin and the heritage of the past all play a role. Above all, altitude effects farming.

Over a quarter of Wicklow's land area lies at over 300 metres above sea-level. Land at this height is several degrees colder than low land all year round. It has a much higher rainfall than the lowlands and is bleak and wind-swept. In winter, frost is almost a permanent feature and heavy snow-falls are common. Settlement is normally not found at this altitude in Ireland. It is difficult to grow crops and the mountain sides are covered in heather and coarse grass, and so are unsuitable for cattle grazing.

During the 17th century the population of Ireland was much greater than it is today, this led to great pressure on the land. In Wicklow, people began to move up on to the mountain sides, and struggled to grow enough food on the acid soils. After the Great Famine the population dropped and the pressure on the land decreased. Many of the mountain farms and huts were abandoned, and can still be seen in isolated areas. In several parts of Wicklow, 'booley' huts were used by farmers as temporary homes during the summer months, when the cattle were taken up to higher slopes for pasture. This vertical movement, up and down the mountain side is still found in many mountainous parts of Europe, such as Norway and Switzerland. The flocks of sheep are moved up on to the mountains in summer and are brought down to graze on the lowlands in the winter.

Traditional sheep farming survives in highland Wicklow today. The farmers live in the glens and each has a patch of land (usually about 5 hectares) on the valley floor. In addition, huge tracts of mountain side are held in common ownership. Each farmer has the right to graze a number of sheep on the unfenced mountain side. Most farmers keep a cow or two and grow potatoes, cabbage and turnips for their

own use, but sheep are their real business. The sheep farmer's year begins with the lambing. He will frequently stay up all night long caring for the ewes and the newborn lambs. Winter in Wicklow is long and severe, with heavy snows even in the glens. The farmer often has to rescue sheep buried in high drifts. In spring he dips the sheep, innoculates them and 'paints' their feet to prevent foot rot. Before releasing them on the mountain side he will mark them with a dye which is his own brand so that he can identify his sheep.

Sheep are shorn each year, this is skilled job as the fleece must come off in one piece. The wool is packed and marketed locally. In late autumn the sheep are brought down from the mountains and their winter grazing may be supplemented by a root crop such as turnips. Sheep-meat prices have increased since Ireland's entry to the EEC, and since 1977, good quality lamb may be sold on the French market where it can fetch very high prices. Wool prices are good also, and for the first time the future of the Wicklow sheep farmer looks bright. He now has the possibility of making a comfortable living but he still has to work hard. The sheep-farmer's only aid is his sheep-dog. The Wicklow sheep dogs are a local breed, the Wicklow Collie, renowned for their intelligence and stamina and their sixth sense for dealing with sheep.

In constrast to the traditional sheep farms are the new intensive factory-type farms which are being developed to supply food for the Dublin market, for example Ballyfree farms in Glenealy. Tens of thousands of chickens are reared indoors, laying thousands of dozens of eggs each week. Turkeys are also reared in battery units and are slaughtered, frozen and packed on the farm. Beef production is carried on along similar lines in the western lowland areas. In the southern part of lowland Wicklow tillage is common, as the climate is drier, warmer and sunnier. Barley, sugar beet and vegetables are grown.

MINING

Mining for various mineral ores has been carried on in County Wicklow over a long period of history. There is evidence that ancient peoples mined copper, and perhaps gold, in the area. Large-scale extraction has been taking place for almost 250 years.

Lead Mines: Where the Wicklow Granite meets with outcrops of slate, lead ore was formed. Lead mines were worked at Luganure and Glenmalure, Co. Wicklow, throughout the nineteenth century. The Luganure mines were worked by "*The Mining Company of Ireland*" which extracted up to 1000 tons of ore each year, valued at up to £12 per ton, a huge sum of money in the middle of the last century. The ore was dragged out of the mine by mules, each pulling three wagons on a small iron tramway. It was then hauled to the railway station at Rathdrum. The Ballinafunshoge mines in Glenmalure were in operation by 1750 and the ruins of the mine buildings and a small smelter can still be seen in the valley today.

The Avoca Mines: The presence of copper ore in the Vale of Avoca was first discovered about 1730 and documents show that mining was in progress in 1753. At that time, the ore was being smelted on the site. By the nineteenth centruy, several different companies were operating mineshafts in the Avoca area. One of them, the *Wicklow Mining Company* built Arklow Harbour for exporting the ore to smelters at Swansea. Wicklow ore was also exported to Liverpool, from about 1830 onwards. The ore was hauled to Arklow by cart from most mineshafts, but the owner of one shaft, the Ballygahan mine, built 12 kilometres of railway track to connect his mine with the harbour.

In the 1830's, pyrite became the main ore mined. Good quality pyrite occurred in huge quantities beside the copper. Pyrite was in huge demand in Britain at that time, as a source of sulphur for industry. Italy was Britain's major supplier but it was feared that in wartime supplies would be cut off. Wicklow was the only place in the British Isles with the potential to supply British Industry. Over 100,000 tons of ore were being extracted each year at Avoca and 800 people were employed, 400 of them as underground miners. In 1836, the Ballymurtagh shaft at Avoca produced £19,943 worth of pyrite, in 1852 this figure had increased to £41,270.

The site of the copper mines at Avoca. In the foreground can be seen the stores of pyrite, the most profitable product of the mine. These supplies of pyrite are bought by the NET fertilizer factory near Arklow. In the centre are the main buildings and the actual pit-face of the mines.

In this same year the production of each type of ore in this shaft was as follows:

Copper Ore	2,233 tons
Copper Pyrites	4,058 tons
Pyrites	24,472 tons

Four other mines were operating at this time—Ballygahan, Cronbane, Tyrony and Connaree. Conaree was making £10,000 profit each year; up to 12 ounces of silver and ½ ounce of gold were being extracted from each ton of ore as well as valuable copper yields. Steam engines were used to power the lifting shafts and winders at Ballymurtagh, but water power was used to provide energy for the machinery at Ballygahan.

Mining is a difficult and dangerous business and the Avoca mines had their moments of excitement. In 1835 the Ballymurtagh mine had a huge rockfall—30,000 tons of ore collapsed at a depth of 33 metres. Eighteen miners were trapped by the fall, and were dug out alive after several days only because a disused shaft was discovered which bypassed the rock-fall. On March 17th, 1845, another huge slump occurred at Ballymurtagh. The mine collapsed downwards from the surface ruining the shafts and the mine buildings. Fortunately nobody was killed because the mine was closed to celebrate St. Patrick's Day. A similar surface collapse occurred at Ballygahan in 1850. A house on the surface was destroyed and a child was killed.

Mining continued at Avoca until 1888, when the mines were closed because of competition from new sources of cheap ore in other countries. They remained closed until World War II, when Irish Industry was short of sulphur, and in 1940 the semi-state mining company *Mianraí Teoranta* began to explore the old Avoca mine shafts. Low grade ore was discovered, and in 1955 *Mogul Mining Corporation of Canada* were invited to take over the lease for 21 years. As an encouragement, the company was granted tax exemption for seven years. Avoca mines began to produce ore once again in 1958, but the company ceased operations in 1962 when the tax concession ran out. The company claimed that low quality ore, poor copper prices and mining difficulties had caused it to lose over £2 million in four years. In 1969 Avoca Mines Ltd., a Canadian financed company, took over the lease and production began, yet again, 1971. The copper ore is now concentrated and shipped from Arklow port to Spain to be smelted. Pyrite is bought by the Nitrigin Eireann Teoranta (NET) factory at Arklow for use in the manufacture of fertiliser. The workforce gradually increased, and there were 400 people employed at the peak, in 1974. World copper prices fell in that year and the number employed dropped to 180. Conditions improved once more, and the workforce later increased to 215 people.

The future of the Avoca mines remains uncertain. The ores are low grade and are really profitable only when world copper prices are high. Mineral prices on the world market are very unsteady. Third World countries which depend on mineral ore exports are at the mercy of a slight drop in prices in New York or London. Avoca is subject to the same pressures, on a smaller scale. There is at least enough ore present for a further 30 years production but it is impossible to predict whether or not the mine will remain open to extract it.

Wicklow Gold

The Aughrim River, which joins the Avoca at Woodenbridge, is also known as the "Gold Mines River". In 1796, a gold nugget was discovered in the river and a 'gold rush' resulted. In 1798, the Government became interested and sent in the army to clear the prospectors out of the area. The army claimed to be searching for rebels, but much of its time was spent in trying to locate the source of the gold dust which turned up occasionally in the river mud. Eventually, the army moved out and the gold panners moved back in but their efforts brought them little return. Altogether about 250 kilogrammes of gold was recovered from the river. The real "bonanza", if it exists, remains undiscovered to this day.

FORESTRY

Until the 13th century, Wicklow was a densely wooded area, with forests of native oak, birch and mountain ash in the valleys and on the lower slopes. The Normans began to cut down the forests to provide timber for castle and ship building, and staves for barrel making. Later, wood was used for smelting ore, and after the 1798 rebellion, the remaining Wicklow forests were cleared to root out the rebels from the Glens of Wicklow. Today, the last remains of the old woods are in Shillelagh. Wicklow oakwoods in the past provided the timber for St. Patrick's

Cathedral, Dublin, and, it is said, for Westminster Abbey in London.

The Irish Government recognised that afforestation was necessary and some of the first of the new forests were planted in Wicklow in the early 1920's. There were many good reasons for planting forests. Hilly land with poor soils was of little use for farming, but ideal for forestry; planting and harvesting trees would provide badly-needed employment; it would not longer be necessary to import all of the wood needed; new forests would provide shelter on bleak mountain sides and so help the farmer by preventing soil erosion. Forests therefore, would be a welcome amenity. One thousand hectares were planted in the Glen of Imaal in 1925, and 1,100 hectares in the Blessington area 1927. There have been a number of further state plantations since then, including those at Saggart (750 h.), Hollywood (550 h.) and Glencree (1400 h.). Altogether, State forests now occupy 8,000 hectares of ground in County Wicklow, mainly in West Wicklow.

Deciduous forests take up to 150 years to mature and so are unprofitable as a 'crop'. Coniferous forests may reach maturity in as little as 50 years and so varieties such as Norway Spruce, Scots Pine, Japanese Larch and Douglas Fir have been planted. When the Wicklow forests have fully matured they will provide raw material for the paper and textile industries. Even now, 'thinnings' can be made every five years. These thinnings can be used for Christmas trees, fencing and telegraph poles and are also used as a raw material

for the manufacturing of chipboard at Athy and Kildare, and of paper at Clondalkin.

The Forestry Division has laid out adventure trails and picnic spots and have opened forests as a recreational facility. The new forests also provide a nature reserve for many forms of wild life. The badger, stoat, squirrel and deer and birds such as the kestrel, sparrow hawk, robin and wren have found refuge in the newly-created woods.

The Life Cycle of a Wicklow Forest

When the site for a forest has been chosen and acquired, the land is fenced and ploughed to provide good drainage and firm rooting bed for the young trees. Seeds are selected from healthy trees and sown in special beds at the forestry research centre at Avondale, near Rathdrum. In 3-4 years, their roots will have developed and they are planted out in the new forest area. For the next two years, the forest needs a lot of care: the young trees must be sprayed to prevent disease and insect attack, and weeds must be controlled to leave room for growth. After about fifteen years, the forest is thinned, and the less successful trees are removed. Lower branches are pruned to promote the growth of a straight, knot-free trunk. The thinnings are sent to a mill for processing as chipboard, wallboard of paper pulp. Every five years thereafter the forest is thinned; by the time the forest is 25 years old, the thinnings are suitable for fencing posts, poles and rough lumber. After 50 years the trees will be mature and will be harvested for use as telephone/electricity poles, and for hundreds of uses in the building industry.

BUILDING STONE

Wicklow Granite is an excellent building material. It is hard to quarry and to cut, and heavy to transport, but it is more durable than other building stones, such as limestones, which are easily eroded by the weather. Granite is very attractive with tiny shiny crystals of mica and other minerals in the stone. In the past, granite was widely used in Dublin city. Large public buildings such as the Custom House, Four Courts, GPO and the Bank of Ireland, were built using massive blocks of granite. The basements and front steps of houses in the Georgian squares were faced with granite. Hundreds of small cottages in South Dublin were built of granite, as were the mansions of wealthy landowners. Much of this stone was quarried at Dalkey, Co. Dublin, and at a number of locations in Wicklow, especially Ballyknockan, near Blessington. This is a small village on the eastern shore of the lake, with a long tradition of quarrying and stone-cutting.

Nowadays concrete blocks, clay bricks and steel reinforced concrete are the main building materials used in Dublin. Granite blocks are expensive to quarry and to lay. New buildings are being faced with granite slabs, however, to give them a solid-looking, attractive frontage. It is sad to reflect that much of this granite is being imported from Sweden. Wicklow granite is no longer quarried in sufficient quantity to meet modern building needs.

Sand and Gravel

Dublin is expanding rapidly, with new houses and hundreds of other buildings being built each year. An enormous quantity of sand and gravel is needed for block making, plastering, ready-mixed and pre-cast concrete, and for road making.

This sand and gravel must be plentiful and free of silt and clay. It must be near the surface of the ground and located close to the city in order to keep down the cost of extracting and transporting it. The sand and gravel ridges of West Wicklow, laid down at the end of the Ice Age by retreating ice sheets, provide high quality raw material for all building purposes. Huge open-cast gravel pits are operated in the Blessington and Ballymore Eustace areas by Cement-Roadstone Ltd. and Braithwaite (Ireland) Ltd.

Unfortunately gravel pits are very unsightly and can permanently damage scenery if they are not landscaped properly. They also create a lot of dust, and are extremely noisy in operation.

PEAT

Blanket bog, up to 2.5 metres deep, has formed on the higher parts of the Wicklow Mountains, mainly due to the high rainfall. These patches of bog are isolated and cannot be worked by machinery like the deep, extensive lowland bogs of the centre of Ireland. They provide peat banks where Dubliners can cut a supply of turf for winter use. These banks are worked in the high Featherbed area, near Kippure Mountain, and are easily seen from the Rathfarnham-Glencree road.

A GREEN AND PLEASANT LAND?

Wicklow is rich in resources which makes it an ideal area for holidays, sightseeing and recreational activities.

A cut away peat bog.

The scenery of the county is among the finest in Ireland, with a wide variety of landscape features, such as mountain, lowland and valley; sea, lake and river; beach, cliffs and forest. The highland is dotted with remote, unspoilt glens, oases of green amid the bleak splendour of the bare brown granite domes.

The roads of Wicklow are of good quality and give easy access to the seaside resorts of the East and the Blessington Lakes in the West. The highland is crossed by the Military Road, allowing the motorist to penetrate even the most isolated areas. At the same time, it is easy for the hillwalker and camper to avoid traffic and main roads.

Rivers, lakes and sea provide facilities for bathing, canoeing, fishing, sailing and water-skiing, and there are many safe sandy beaches where children can play safely. Forests are well laid-out with nature walks and picnic tables. Historic sites, such as St. Kevin's Monastery at Glendalough and Baltinglass Abbey attract thousands of visitors each year. Accommodation is readily available in youth hostels, farm houses, caravan sites and hotels.

In recent years, Wicklow has become a major centre for adventure sports: it has many features which are ideal for activities of this kind—whitewater rivers for the canoeist, cliffs and rock-faces for the climber, forests for the orienteer and the open mountain for the hillwalker.

Nowadays, more and more people have leisure time and money to enjoy it. Dubliners are fortunate in that they can visit even the remotest areas of Wicklow on day trips. More than ever, there is a need to conserve the beauty and peace of rural Wicklow and to ensure that its landscape is not spoiled and polluted by those who use it.

All of Wicklow's resources are exploited in one form or another, and have been exploited for centuries past. In some cases this involves millions of pounds' worth of technology, as in the Avoca underground mines. But resources may also be exploited very simply by human labour alone, as in the turf cuttings at Glencree.

The natural wealth of Wicklow must be carefully managed so that its beauty is not destroyed. The deposit heaps of Avoca mines must not be allowed to destroy the beautiful Vale of Avoca. The new forests will be a great aid in soil conservation and may provide a preserve for wild life. This is a good example of environmental planning.

The resources of the earth are delicately balanced and interdependent. The use of any resource must be carefully planned in order that it may be conserved, and also in order that its use may not do permanent damage to any other resource or to the whole environment.

END